EAI/Springer Innovations in Communication and Computing

Series editor
Imrich Chlamtac, European Alliance for Innovation, Ghent, Belgium

Editor's Note

The impact of information technologies is creating a new world yet not fully understood. The extent and speed of economic, life style and social changes already perceived in everyday life is hard to estimate without understanding the technological driving forces behind it. This series presents contributed volumes featuring the latest research and development in the various information engineering technologies that play a key role in this process.

The range of topics, focusing primarily on communications and computing engineering include, but are not limited to, wireless networks; mobile communication; design and learning; gaming; interaction; e-health and pervasive healthcare; energy management; smart grids; internet of things; cognitive radio networks; computation; cloud computing; ubiquitous connectivity, and in mode general smart living, smart cities, Internet of Things and more. The series publishes a combination of expanded papers selected from hosted and sponsored European Alliance for Innovation (EAI) conferences that present cutting edge, global research as well as provide new perspectives on traditional related engineering fields. This content, complemented with open calls for contribution of book titles and individual chapters, together maintain Springer's and EAI's high standards of academic excellence. The audience for the books consists of researchers, industry professionals, advanced level students as well as practitioners in related fields of activity include information and communication specialists, security experts, economists, urban planners, doctors, and in general representatives in all those walks of life affected ad contributing to the information revolution.

Indexing: This series is indexed in Scopus, Ei Compendex, and zbMATH.

About EAI

EAI is a grassroots member organization initiated through cooperation between businesses, public, private and government organizations to address the global challenges of Europe's future competitiveness and link the European Research community with its counterparts around the globe. EAI reaches out to hundreds of thousands of individual subscribers on all continents and collaborates with an institutional member base including Fortune 500 companies, government organizations, and educational institutions, provide a free research and innovation platform.

Through its open free membership model EAI promotes a new research and innovation culture based on collaboration, connectivity and recognition of excellence by community.

More information about this series at http://www.springer.com/series/15427

Satish R. Jondhale • R. Maheswar • Jaime Lloret

Received Signal Strength Based Target Localization and Tracking Using Wireless Sensor Networks

Springer

EAI
RESEARCH MEETS INNOVATION

Satish R. Jondhale
Department of Electronics and Telecommunication
Amrutvahini College of Engineering
Sangamner, Maharashtra, India

R. Maheswar
Research
VIT Bhopal University
Bhopal, Madhya Pradesh, India

Jaime Lloret
Instituto de Investigación para la gestión Integrada de Zonas Costeras
Universitat Politecnica de Valencia
Valencia, Valencia, Spain

ISSN 2522-8595 ISSN 2522-8609 (electronic)
EAI/Springer Innovations in Communication and Computing
ISBN 978-3-030-74060-3 ISBN 978-3-030-74061-0 (eBook)
https://doi.org/10.1007/978-3-030-74061-0

This Springer imprint is published by the registered company Springer Nature Switzerland AG
The registered company address is: Gewerbestrasse 11, 6330 Cham, Switzerland

Preface

Location awareness is a key component in many industrial, scientific, and military indoor and outdoor applications as well as a wide variety of present location-based services (LBS). Although GPS is a more popular technique to get location updates very easily, limited accessibility GPS signals in most of the indoor as well as outdoor environmental setup motivate researchers to design GPS-less localization system. Being one of the key technologies of twenty-first century, the low powered and low cost wireless sensor network (WSN) paved the way for the design and development of GPS-less system for indoor as well as outdoor localization and tracking (L&T) applications. The merits of the WSN technology over the rest of the other technological alternatives are: easy deployment, small size, low cost, low power consumption, and ad hoc nature. Due to no additional hardware requirement and simplicity in the usage, the received signal strength indicator (RSSI) is the most widely used metric of field measurement in WSN-based L&T systems as compared with other possible metrics. However, the existing RSSI-based target tracking systems generally suffer with low tracking accuracy because of signal propagation issues such as reflection, refraction, multipath propagation, and non-line of sight (NLOS). Apart from signal propagation issues, environmental dynamicity aspects such as abrupt variations in target velocity during motion, nonavailability of all RSSI measurements, variations in target mobility patterns also make RSSI-based target L&T highly challenging. Although much research has already been done in WSN-based L&T, most of these existing systems are not robust and efficient in terms of tracking accuracy and computational complexity. The present focus of all the researchers working in RSSI and WSN-based L&T domain is the development of efficient, robust, and accurate L&T system. The research in WSN- and RSSI-based L&T domain is blooming with very high pace that it is very difficult to encompass all the new developments in it; however, we tried our best to provide a detailed review of recent and relevant information of existing RSSI- and WSN-based L&T systems. The main focus of writing this book is to give a systematic approach of learning fundamentals of WSN and its capability to build L&T applications. The sincere attempt is made in this book to answer about how to design novel-efficient RSSI-based tracking system which can track single mobile target and yield high tracking accuracy irrespective of its motion. Several artificial neural network (ANN)-based implementations dealing with tracking of single mobile target with environmental dynamicity are presented in this book and are validated through extensive MATLAB-based simulation experiments. We believe that this book can provide an effective way to design or program customized solution tailored to meet the underlying WSN-based L&T applications with the help of RSSI measurements.

Thus through this book, we not only present the fundamentals of RF communication, WSN-based target L&T, hardware, protocols architectures, and pros-cons in the existing RSSI- and WSN-based systems, but we also present system-level implementation through MATLAB-based building blocks of subsystems of L&T system. One can use these ready-to-use building blocks to understand and build their WSN-based L&T applications or pursue further research to customize their underlying application as per the actual requirement. Any undergraduate student of physics, mathematics, computer science, or electronics disciplines might feel comfortable to follow this book material.

Sangamner, India — Satish R. Jondhale
Bhopal, India — R. Maheswar
Valencia, Spain — Jaime Lloret

Acknowledgment

I would like to express my sincere thanks to Prof. Chlamtac (President, European Alliance for Innovation (EAI)) and Eliška Vlčková (Managing Editor, EAI) for providing the opportunity to write the book entitled, *Received Signal Strength Based Target Localization and Tracking Using Wireless Sensor Network*. As a corresponding author for this book, I would like to give special thanks to the book co-authors Dr. R. Maheswar (School of EEE, VIT Bhopal University, Bhopal) and Dr. Jaime Lloret (Polytechnic University of Valencia, Spain) for being selfless mentors, brilliant research partners, and precious friends during the overall accomplishment of this book. I would like to thank Dr. Rajkumar S. Deshpande (my Ph.D. guide), Dr. D. N. Kyatanvar (Principal, Sanjivani COE, Kopargaon), and Dr. B. S. Agarkar (Sanjivani COE, Kopargaon) for motivating me to extend my Ph.D. work in the context of writing this book. I also thank the Management, Dr. R. P. Labade (Head, E&TC department), and my colleagues from Amrutvahini COE, Sangamner, Maharashtra, India for giving me all kind of support and facilities to complete this book successfully. I also thank all the reviewers for giving their precious feedback to improve the work further. I would also like to thank all the supporting staff from Springer who really helped a lot and their extended support with quick and efficient efforts made the book finally a successful one. Sincere thanks to my wife Prof. Amruta, my daughters Aarohi and Rajlaxmi, and my parents for wholeheartedly excusing my absence during precious life moments when I was writing this book. I feel that without the support of my family members, this book writing would not at all be possible. At the end, I must extend a huge expression of gratitude to Lord Shri Krishna for offering me enough energy and knowledge during the making of this book.

Contents

About the Editors

Satish R. Jondhale received his B.E. in Electronics and Telecommunication in 2006, his M.E. in Electronics and Telecommunication in 2012, and his Ph.D. in Electronics and Telecommunication in 2019 from Savitribai Phule Pune University, Pune, India. He has been working as an Assistant Professor in Electronics and Telecommunication Department at Amrutvahini College of Engineering, Sangamner, Maharashtra, India for more than a decade now. His research interests are Signal Processing, Target Localization and Tracking, Wireless Sensor Networks, Artificial Neural Networks and Applications, Image Processing and Embedded System Design. He has several research publications in reputed journals such as IEEE Sensors Journal, Ad Hoc Networks (Elsevier), Ad Hoc & Sensor Wireless Networks, and International Journal of Communication Systems (Wiley). He has published two book chapters in *Handbook of Wireless Sensor Networks: Issues and Challenges in Current Scenario*, Springer, 2019. He is also a member of professional societies such as IEEE and ISTE. He has been a reviewer for peer-reviewed journals such as *IEEE Transactions on Industrial Informatics*, *IEEE Sensors*, *Signal Processing* (Elsevier), *IEEE Access*, *IEEE Signal Processing Letters*, and *Ad Hoc & Sensor Wireless Networks*, and so on. He has received review recognition appreciation from Mississippi State University, USA for valuable review work. He had served as a TPC member for Sixth International conference on Internet of Things: Systems, Management and Security (IOTSMS, 2019) held at Granada, Spain from 22 to 25 October, 2019 (Technically Co Sponsored by IEEE Spain Section). He has been appointed as "Bentham Brand Ambassador" for 2019–20.

R. Maheswar has completed his B.E (ECE) from Madras University in the year 1999, M.E (Applied Electronics) from Bharathiar University in the year 2002 and Ph.D. in the field of Wireless Sensor Network from Anna University in the year 2012. He has about 19 years of teaching experience at various levels and presently working as Dean–Research (Assistant) and Dean In-Charge for the School of EEE, VIT Bhopal University, Bhopal. He has published around 70 papers at International Journals and International Conferences and published 4 patents. His research interest includes Wireless Sensor Network, IoT, Queueing theory, and Performance Evaluation. He has served as guest editor for Wireless Networks Journal, Springer and is serving as editorial review board member for peer-reviewed journals, and also edited four books supported by EAI/ Springer Innovations in Communications and Computing book series. He is presently an associate editor in Wireless Networks Journal, Springer, Alexandria Engineering Journal, Elsevier and Ad Hoc & Sensor Wireless Networks Journal, Old City Publishing.

Jaime Lloret received his B.Sc.+M.Sc. in Physics in 1997, his B.Sc.+M.Sc. in electronic Engineering in 2003 and his Ph.D. in telecommunication engineering (Dr. Ing.) in 2006. He is a Cisco Certified Network Professional Instructor. He is IEEE Senior, ACM Senior, and IARIA Fellow. He is Chair of the Integrated Management Coastal Research Institute (IGIC), IEEE Spain Section Officer, Chair of the Internet Technical Committee (IEEE Communications Society Internet Society) (Term 2014–2015), Head of the Innovation Group "Active and collaborative techniques and use of technologic resources in the education (EITACURTE)" as well as Chair IEEE 1907.1 WG (till 2018). He is currently Associate Professor in the Polytechnic University of Valencia. He is the Chair of the Integrated Management Coastal Research Institute (IGIC) and he is the head of the "Active and collaborative techniques and use of technologic resources in the education (EITACURTE)" Innovation Group. He is the director of the University Diploma "Redes y Comunicaciones de Ordenadores" and he has been the director of the University Master "Digital Post Production" for the term 2012–2016. He was Vice-chair for the Europe/Africa Region of Cognitive Networks Technical Committee (IEEE Communications Society) for the term 2010–2012 and Vice-chair of the Internet Technical Committee (IEEE Communications Society and Internet society) for the term 2011–2013. He has been Internet Technical Committee chair (IEEE Communications Society and Internet society) for the term 2013–2015. He has authored 22 book chapters and has more than 480 research papers published in national and international conferences, international journals (more than 230 with ISI Thomson JCR). He has been the co-editor of 40 conference proceedings and guest editor of several international books and journals. He is editor-in-chief of *Ad Hoc and Sensor Wireless Networks* (with ISI Thomson Impact Factor), the international journal *Networks Protocols and Algorithms*, and the *International Journal of Multimedia Communications*. Moreover, he is Associate Editor-in-Chief of *Sensors* in the Section Sensor Networks, he is advisory board member of the *International Journal of Distributed Sensor Networks* (both with ISI Thomson Impact Factor), and he is IARIA Journals Board Chair (8 Journals). Furthermore, he is (or has been) associate editor of 46 international journals (16 of them with ISI Thomson Impact Factor). He has been involved in more than 450 Program committees of international conferences and more than 150 organization and steering committees. He has led many local, regional, national, and European projects. He is currently the chair of the Working Group of the Standard IEEE 1907.1. Since 2016, he is the Spanish researcher with highest h-index in the TELECOMMUNICATIONS journal list according to Clarivate Analytics Ranking. He has been general chair (or co-chair) of 52 International workshops and conferences (chairman of SENSORCOMM 2007, UBICOMM 2008, ICNS 2009, ICWMC 2010, eKNOW 2012, SERVICE COMPUTATION 2013, COGNITIVE 2013, ADAPTIVE 2013, 12th AICT 2016, 11th ICIMP 2016, 3rd GREENETS 2016, 13th IWCMC 2017, 10th WMNC 2017, 18th ICN 2019, 14th ICDT 2019, 12th CTRQ 2019, 12th ICSNC 2019, 8th INNOV 2019, 14th ICDS 2020, 5th ALLSENSORS 2020, Industrial IoT 2020 and GC-ElecEng 2020, and co-chairman of ICAS 2009, INTERNET 2010, MARSS 2011, IEEE MASS 2011, SCPA 2011, ICDS 2012, 2nd IEEE SCPA 2012, GreeNets 2012, 3rd IEEE SCPA 2013, SSPA 2013, AdHocNow 2014, MARSS 2014, SSPA 2014, IEEE CCAN 2015, 4th IEEE SCPA 2015, IEEE SCAN 2015, ICACCI 2015, SDRANCAN 2015, FMEC 2016, 2nd FMEC 2017, 5th SCPA 2017, XIII JITEL 2017, 3rd SDS 2018, 5th IoTSMS 2018, 4th FMEC 2019, 10th International Symposium on Ambient Intelligence 2019, 6th SNAMS 2019, and ACN 2019, and local chair of MIC-WCMC 2013 and IEEE Sensors 2014).

Chapter 1
Fundamentals of Wireless Sensor Networks

1.1 Introduction to Wireless Sensor Network

The WSN can be described as autonomous and self-organizing systems that consist of a large number of tiny, low-cost, battery-operated sensor nodes (also called as motes), which are generally randomly deployed either inside the phenomenon of interest or very close to it [1, 2]. These motes are generally utilized to monitor environmental and physical conditions, such as pressure, temperature, light, humidity, fire detection, and chemical level [3–5]. These nodes can sense the environment (data collection) and process and forward the processed data directly to the base station (also called as sink) or via the other sensor nodes to the base station in order to process it further as per application requirements. These sensor nodes in the WSN are fitted with an onboard processor. Instead of forwarding the raw sensed data, sensor nodes use their built-in processing capability to carry out simple computations at local level and then transmit only the partially processed data to the nodes responsible for the fusion with that obtained from other nodes. These computational capabilities in WSN ensure a wide range of applications [3] [6–8]. For example, it is possible to monitor the physiological data of a patient remotely by a doctor, which saves a lot of time of patients as well as doctors. The WSN can also be used to locate and/or detect pollution level as well as percent of toxic contents in the air and the water. Thus, the WSN can provide the end user a better understanding of the environment with intelligence. In 10–15 years it is not unreasonable to expect that the large portion of the world will be covered with WSN with access to them via Internet [3, 6].

A typical WSN model consisting of sensor nodes, sink, Internet connectivity, and end user is shown in Fig. 1.1. Sensor field is nothing but an environment under consideration, wherein the nodes are deployed to gather the information in it [5]. Each of the nodes is capable of sensing, processing, and forwarding the sensed data to the

S. R. Jondhale et al., *Received Signal Strength Based Target Localization and Tracking Using Wireless Sensor Networks*, EAI/Springer Innovations in Communication and Computing, https://doi.org/10.1007/978-3-030-74061-0_1

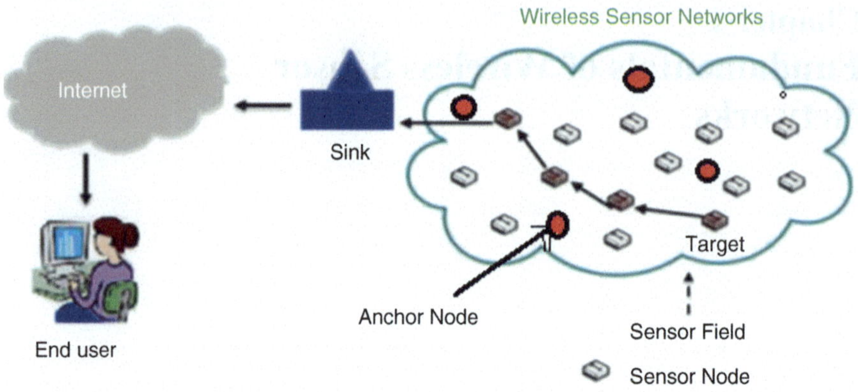

Fig. 1.1 A typical WSN model

requested nodes or to the sink. The sink and the sensor node may be static or mobile, depending upon the application requirements. The sink can collect as well as process the data from the sensor nodes. Generally, the sink is rich in memory, computational capacity, and energy as compared to the sensor nodes. The sink connects the WSN to the end user with a terminal (such as computer) using an existing communication infrastructure such as Internet.

The concept of WSN can be described with the help of the following simple equation [3, 5]:

$$\text{Sensing} + \text{CPU} + \text{Radio} = \text{Thousands of possible applications of WSN}$$

Thus, knowing the capabilities of the WSN, thousands of applications appear in mind. Although it seems a straightforward combination of modern technologies, combining sensors, processors, and radios into a coin-sized node requires a detailed knowledge of both the capabilities and limitations of each of the underlying hardware part, as well as fundamentals of distributed systems and modern networking technologies [3, 5]. Each of the sensor nodes must be designed to encompass the set of primitives required to formulate the network of nodes, while strictly attaining requirements of cost, size, and power consumption. Thus, the major challenge is to map the overall system requirements.

The WSN node may have different types of sensors interfaced, capable to monitor a wide variety of ambient conditions. The type of sensor can be seismic, pressure, thermal, magnetic, infrared, acoustic, and visual [3, 5]. The sensor nodes can be deployed manually or randomly. Although each individual node has several resource constraints in terms of memory, energy, computation, and communication capabilities, their heavy deployment can collectively sense the surrounding environment, disseminate measurements, and process these experimental measurements. That's why the WSN applications range from environmental monitoring, real-time tracking, to structural health of monitoring [3, 6–8]. Capability to real-time information makes the WSN an ideal candidate for handling emergency, disaster relief operations, and military that need efficient coordination and planning. The WSN can also be useful for instrumenting and controlling of offices, factories, vehicles, homes, and cities. Any WSN-based application is useful only if the location of the sensor node that provides the measurements is correctly known. In other words, node localization is of prime importance to any WSN-based application [9]–[12]. In order to get the node locations, an effective localization algorithm is needed. The WSN-based system gets the updates of the location of node that provide useful measurements; however, many times the estimated locations are not trustworthy because of noisy measurements. Thus, in most of the situations, location estimates are not accurate enough to claim that the underlying WSN-based application is robust and reliable. That's why the node localization has attracted tremendous attention of the researchers. The major objective of any localization algorithm is to improve node localization accuracy (i.e., to reduce localization error). In recent years, one of the major researches in WSN domain is on localization and tracking (L&T). Designing efficient L&T algorithms becomes an important factor for the success of any WSN-based application [9–12].

In this book we provide the fundamental aspects of WSN as well as a detailed framework of WSN-based L&T system right from concept to design. We cover fundamentals of RSSI-based L&T using WSN, simulated as well as real-time WSN-based L&T framework. A sincere attempt is made to provide the survey of the existing RSSI-based L&T systems through a rigorous review of literature from the recent papers of journals as well as conferences. This book is targeted to the managers, communications developers, and practitioners, who wish to acquire the knowledge of target L&T and wish to implement WSN-based L&T system to encompass broad range of related applications.

1.2 WSN Versus Other Wireless Networks

The advancements in RF domain and rise in portable devices have accelerated the use of mobile and wireless networking [13–15]. Because of wireless networking, the users can electronically access data and services, irrespective of their physical location [8]. Wireless technology-based networks are generally classified into two categories, namely, infrastructure-based networks and infrastructure-less networks (ad hoc networks). The former category has fixed the base station called access points, which are connected by wires. The mobile node can communicate with the base station via wireless link if it is inside the communication range of that base station. If this mobile node travels out of the communication range of that base station, then it tries to establish the connection with the other base station inside whose communication range it currently is. Cellular phone system, paging systems, and wireless local area networks (WLAN) are some of the examples of infrastructure-based networks, whereas ad hoc networks do not have such predefined infrastructure and the nodes can move freely from one place to another, changing the network topology continuously [3–5]. Mobile ad hoc network (MANET) and WSN are some of the examples of ad hoc networks. These networks do not necessitate previous setup or supporting infrastructure.

The MANET is a network of self-configurable, autonomous, self-organizing nodes with wireless communication capabilities (especially multi-hop communication) [16]. It is generally adopted to meet the requirements of immediate communication need, where the deployment of wired infrastructure is not a feasible option. For instance, MANET is used in situations such as battlefield, disaster relief operations, flood relief operations, and large construction sites. As compared to wired infrastructure, the MANET can cover larger geographical areas. The WSN is a special kind of ad hoc network, consisting of heavily deployed sensor nodes that can cover a much wider geographical area as compared to the MANET [16]. As described in the previous section, the sensor nodes in WSN are battery operated, low cost, and small in size.

Some of the similarities between the WSN and MANET are:

- Both are distributed wireless networks with no requirement of previous infrastructural setup.
- Nodes are deployed in an ad hoc manner in both.
- In most of the applications, the nodes communicate with each other using multi-hop way.
- Both have concern over the minimization of power consumption due to use of battery-powered nodes.
- Due to uses of unlicensed spectrum for operation, both are generally prone to interference by other RF-based devices operating in the same frequency slot.
- Self-configuration is a must in both due to distributed nature.

In spite of many similarities between the WSN and the MANET, there are also few key differences between them as listed below [3, 5, 6, 16]:

- The node in the WSN is generally of the order of several hundreds to thousands as compared to the small number in the MANET. Thus, node deployment density in the WSN is very high.
- Nodes in the WSN are prone to failure due to environmental and physical conditions.
- Due to frequent node failures, WSN topology gets updated quite often.
- In most of the situations, the WSN use broadcast communication strategy, whereas the MANET adopts point-to-point networking.
- The scarcity of resources is a common problem in the WSN (which means constraints of energy, computational abilities, and memory).
- The WSN nodes generally do not have global unique identification due to mass (heavy) deployment.
- In majority of the applications, node mobility is comparatively low or nil in the WSN as compared to that in the MANET.
- As compared to the MANET, the data rate in the WSN is very low.

1.3 Sensor Node Architecture

Sensor nodes are designed and consisting of many components than just wireless sensors. As mentioned earlier these sensor nodes can sense the physical parameter of interests, process it, and dispatch the processed data to the base station. A sensor node can be defined in the following way [3–5, 8]:

> A sensor node is a type of transducer that senses one type of energy (field measurements) and converts it into a suitable form (electrical form) for the purpose of data transfer to the other sensor nodes. Furthermore, it possesses the ability to avoid the transmission of the redundant data sensed from the surrounding environment (field measurements).

From a hardware perspective, the sensor nodes are small-scale processing units with a variety of sensors interfaced to it. Typically the field measurements are temperature, noise level, wind pressure, the presence of static or moving objects, received signal strength indicators (RSSI), and so on [3–5, 8]. The type of sensors interfaced with the node depends upon the underlying targeted application. Speaking in more specific words, the sensor node typically has inbuilt processor (to process the physical measurements received from the interfaced sensors), a battery (to power it up), a memory (to store raw sensed or processed data), and a radio or communication unit (for communication with the other nodes or external world). The sensor network's networking and communication abilities can be creatively exploited to deal with specific underlying application. Sensor node architecture with these four functional units is illustrated in Fig. 1.2 as shown below.

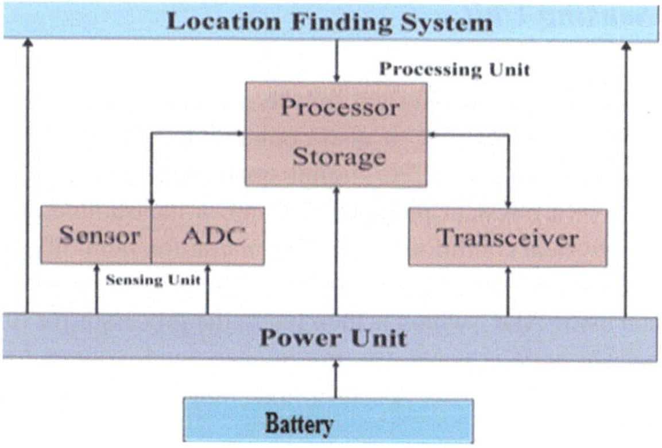

Fig. 1.2 Components of sensor node

1.3.1 The Power Supply

The power supply unit generally includes a nonrenewable coin-sized battery, whose role is to supply power to all of the units of the sensor node [3–5, 8, 10]. Thus, batteries are obviously energy storage devices, whose size ranges from small coin cell to large lead-acid batteries of AA or AAA types. The rechargeable batteries are generally not used in most of the WSN-based applications due to high cost, low energy density, and impracticality of recharging option. If this battery is depleted, the sensor node becomes nonfunctional. As in most of the WSN-based applications, the sensor nodes are deployed in hostile environment and are generally inaccessible; the sensor node lifetime mainly rely on the attached batteries.

In the sensor node, power is consumed for node activities, such as sensing, data processing, and communication. Out of these node activities, the major part of power consumption is observed for data communication. For instance, the power consumption on transmitting 1 Kb data over a distance of say 100 m is approximately the same as that for executing approximately three million instructions by a processor with a capability of 100 million instructions per second (MIPS) [3]. The power consumption is a major design constraint of the WSN due to the limitation in battery size. Thus, designing of power supply unit is a very crucial task in sensor network design for an application. This design part may vary from application to application. However, it is also possible to power the network and extend the WSN lifetime by extracting energy from the environment by the usage of solar cells.

1.3.2 The Sensing Unit

This unit generally consists of physical sensors, which are capable of sensing the physical parameter of interest [3]. It also contains an analog-to-digital converter (ADC) to transform sensed data into digital form. Sensor is a transducer, which converts a change in a physical phenomenon into a measurable electrical signal. Sensors measure physical conditions such as temperature, humidity, light, pressure, sound, chemical level, magnetic fields, and etc. The sensor converts analog signal into digital signal using ADC, which is then fed to the processor for further required processing. A sensor node is generally tiny in shape and requires low power consumption and operates unattended. A sensor node may have several types of sensors connected to the node.

1.3.3 *The Processor Unit*

The processing unit in a WSN node consists of a suitable embedded processor for processing the digital data obtained from ADC unit [3]. The processor can execute various tasks, such as processing of input data and controlling the working of other components of the node. The processing unit generally has a microcontroller to execute all of the mentioned tasks; however, in some of the applications, it may consists of digital signal processor (DSP) and field-programmable gate array (FPGA). The microcontroller is a more preferred option due to low power consumption and low cost involved as well as flexibility of interfacing with other devices and ease in programming. The common microcontrollers that are used in sensor nodes are Atmel ATmega128 series controllers, ARM microcontrollers, Texas Instruments' MSP 430, and Microchip's PIC. The more complex the application, the more advanced microcontroller is preferred in the sensor node to meet the application requirements [3–5, 16].

The processing unit also contains a memory unit for storage of the processed data and algorithms of the underlying application. The memory unit consists of on-chip flash memory, internal RAM, and external flash memory. For instance, Mica2 mote is based on ATmega128L microcontroller, which has 4 Kb static RAM and 128 Kb flash program memory [3].

Though it is the era of modern powerful and tiny processors, the power (energy) and memory of the sensor node are still considered as scarce resources. Some of the typical tasks executed by the processing unit are:

- Control, signal processing, and actuation.
- Data aggregation.
- Compression, clustering, forward error correction, and encryption.
- Data fusion and data analysis

1.3.4 *The Communication Unit*

The communication unit consists of a wireless radio transceiver. For collaborative processing, the sensor nodes frequently need to exchange the data with the neighboring nodes. The transceiver can convert the digital bit stream received from the microcontroller into RF waves or RF waves into an equivalent digital bit stream [3, 16, 17]. Thus, the sensor node can communicate with the external world (other nodes) through interfaced transceiver. The transmission media for communication between nodes can be RF, optical, or infrared. Communications using lasers need less energy; however, they require LOS for communication, and additionally they are sensitive to atmospheric conditions. Like lasers the infrared does not need antenna; however, its broadcasting capacity is limited. The RF communication generally involves various important operations, such as modulation and demodulation, filtering, and multiplexing. These operations make sensor node communication highly complex and expensive as compared to other operations of the sensor node. Additionally, the signal path loss during the communication between two communicating sensor nodes has exponential relation with the distance between them, as the sensor node antennas are usually close to the ground. In spite of the high communication cost involved, the RF-based communication is widely preferred in the WSN-based applications [9–12, 18, 19]. The reason behind this is that the data rates are low and packets are small in RF communication. One more advantage with RF communication is the possibility of frequency reuse due to shorter communication lengths. The transceiver has four operational states, namely, receive, transmit, sleep, and idle. The power consumption in idle mode is almost equal to that in the receive mode. Therefore, if the transceiver is not transmitting or receiving, it's better to shut down it completely rather than leaving it in the idle mode. Another important aspect to note down is that significant power consumption occurs during switching; therefore, unnecessary switching between states needs to avoided. The popular Mica2 mote uses two kinds of RF transceivers, namely, Chipcon CC1000 and RFM TR1000. The transmission range of Mica2 is around 150 m [3]. Some of the dominant wireless standards used for communication by the sensor nodes are:

- IEEE 802.15.1 PAN/Bluetooth
- IEEE 802.15.3/UWB
- IEEE 802.15.4/ZigBee
- IEEE Wi-Fi

1.3.5 Location Finding Unit

As discussed earlier the sensor node positioning is important in any WSN-based application. In WSN locations of few nodes are prefixed (such nodes are called as anchor nodes), whereas the remaining nodes are randomly deployed in the environment and are termed as non-anchor nodes [20–23]. That means locations of the non-anchor nodes are unknown. Since sensor nodes are generally deployed randomly and run unattended, they need to corporate with a location finding system. The location finding unit in the sensor node architecture is optional. If it is present in the sensor node, then it contains a Global Positioning System (GPS) to estimate the location of the node. It is often assumed that each sensor node will have a GPS unit that has approximately 5 m accuracy [24–27]. Equipping all sensor nodes with a GPS is not a viable solution in the WSN due to the cost involved. The possible solution to this is to interface GPS to anchor nodes and then locate the non-anchor nodes with the help of anchor nodes by executing a suitable localization algorithm.

1.4 Sensor Network Communication Architecture

The overall working of the WSN can be explained using the protocol stack as elaborated in Fig. 1.3 [3, 5]. The protocol stack includes five layers, namely, physical layer, network layer, data link layer, transport layer, and application layer. Based on the sensing tasks, a variety of application software may be built and run on the application layer. The transport layer is responsible to maintain the data flow between sensor nodes. The network layer monitors the routing of the data provided by the transport layer. Minimizing the collision with neighbor nodes during broadcast is the main task of the data link layer. The physical layer deals with modulation, data transmission, and data receiving techniques for the WSN.

Apart from these layers, the three management planes associated with the protocol stack are task plane, power plane, and mobility plane (see Fig. 1.3). These three planes monitor the power, movement, and task distribution among the WSN nodes. These three planes assists the sensor node in lowering the overall power consumption and coordinating the sensing task. The power plane takes care of efficient and effective utilization of power among sensor nodes during operation of the network as a whole [3] [5]. For instance, the sensor node turns off its receiver in order to avoid duplication of data. Let's consider another case wherein the power level of a sensor node is low. In such critical situation such sensor node may broadcast to its neighboring nodes that it has low power and can't participate in data routing. In other words, this node will reserve the remaining power only for sensing. The mobility plane is responsible for registering and detecting the movement of sensor

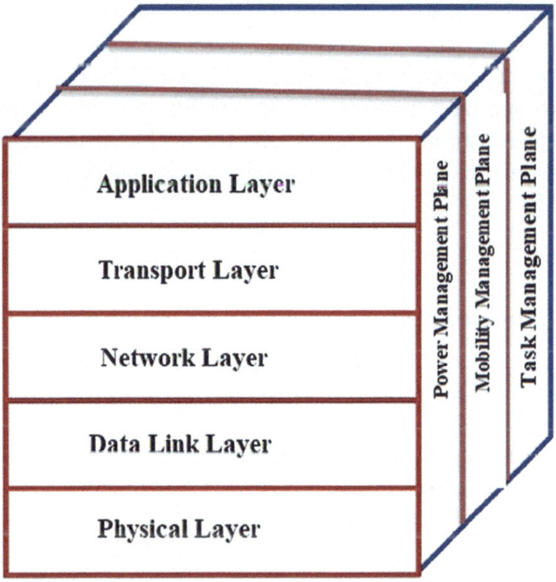

Fig. 1.3 Wireless sensor network protocol stack

nodes. That's why each sensor node can keep a track on the movement of its neighboring nodes. By the knowledge of the neighboring nodes in advance, the sensor nodes can maintain a balance between its task and power usage. The task plane is responsible to balance and schedule the sensing tasks for a specific region in the given monitoring environment. Thus, there is no need for the sensor nodes to sense the environment at the same time. In other words, only those sensor nodes, which have sufficient power level, will perform the sensing tasks. Thus, all of these management planes are essential for the sensor network to route the data effectively in the network, to achieve power efficiency, and to marshal resources among the network nodes [3, 5]. That means without these three management planes, a sensor node could just work individually without a concern about the rest of the network. For the sensor network as a whole, it will be highly advantageous if the sensor nodes in the network can collaborate with each other in order to prolong the lifetime of the sensor networks.

1.5 Design Constraints for WSN

The WSNs are characterized by a very powerful combination of distributed sensing, computing, and communication. Despite the tiny size of an individual WSN node, it faces numerous challenges such as stringent power constraints, limited communication range, computing power, and storage space of the sensor nodes [3–5, 10, 28]. The major reason for these constraints is the small physical size of the sensor nodes. The primary objective of the WSN is to execute the task of data communication (routing) while trying to extend the network lifetime as high as possible by employing energy-efficient techniques. Some other operating challenges include high error rates, low bandwidth, noisy measurements, sleep scheduling of sensor node, scalability to a huge amount of sensor nodes, survivability in dynamic environments, breakdown of wireless communication link, and frequent node failure. The following section discusses some of the important design issues and challenges that affect data routing in WSNs.

1.5.1 Power Consumption

As discussed earlier the sensor nodes are generally battery powered and are generally deployed in remote or inaccessible environments [3, 6–8]. Replacing or recharging the batteries in such environment is almost impossible. The power is a mandatory aspect for almost all of the operations in the WSN. In general, the power consumption in sensor nodes is observed at three places: (a) power consumption by sensing unit, (b) power consumption by communication unit, and (c) power consumption by processing unit. Therefore, the power consumption is one of the major concerns in the WSN-based applications [16, 20, 29, 30]. It is observed that a single bit transmission in the WSN consumes the same power as that for executing approximately 800–1000 instructions. Thus, the power consumption in radio is much higher than that in sensing and computation.

From the architectural point of view, the use of low-power antenna circuitry must be chosen to reduce power consumption. Low power consumption is a key to success in any WSN-based application. That's why a lot of research has been going on in the WSN community to develop energy-efficient algorithms for routing, localization, and other tasks, which will consume less power [16, 20, 29, 30]. Parallel to this, continuous research is going on to extend sensor node lifetime despite its battery-dependent working. Power efficiency in the WSN can be accomplished in three ways:

1. Low-duty-cycle operation.
2. Local/in-network processing to reduce data volume and in turn transmission time.
3. Multi-hop communication reduces the requirement for long-range transmission.
4. Each node in the WSN can act as a repeater, thereby reducing the communication link range coverage.

1.5.2 Memory

The sensor node generally has a very small amount of memory in the processing unit for the storage of data and algorithm [3–5]. This memory is in the form of RAM and ROM of processor of the sensor node. Due to the limited memory capacity of the sensor node, there does not exist enough memory to execute complex algorithms especially after loading the OS. For instance, consider the case of Smart Dust project. In this project it is found that TinyOS consumes around 4 Kb for instructions, leaving only 4.5 Kb for applications [3].

1.5.3 Deployment, Topology, and Coverage

Depending on the application requirement, the nodes in the WSN can be placed in a planned fashion or in a random fashion [20, 31–33]. The node deployment in the monitoring area can be a periodic or a one-time activity. Node deployment has impact on important network parameters, such as coverage, node density, reliability, sensing resolution, communications, and task allocations. The WSN generally operates in dynamic environment due to uncertainty in operating conditions, e.g., due to abrupt changes in the environmental setup, node mobility, and node failures. Due to such dynamicity in the operating environment, the communication links between sensor nodes frequently break even when nodes are static. Another disadvantage of this dynamicity is frequent changes in the WSN topology, which in turn affects many network characteristics such as robustness, latency, and capacity. The level of complexity in data routing and processing also depends on the network topology. Coverage is a measure of coverage area of a WSN. It can be sparse, i.e., only parts of the environment fall under the sensing envelope, or dense, i.e., most parts of the environment are covered. Coverage can also be redundant, i.e., the same physical space is covered by multiple sensors. Coverage is mainly determined by the sensing resolution demands of an application.

1.5.4 Communication and Routing

As the WSN generally has limited bandwidth, processing, and energy, it operates in highly uncertain, remote, and hostile environments [18, 34]. Therefore, the network continuously undergoes changes in its topology and coverage due to frequent node failures and noisy measurements. Due to very heavy deployment, its nodes lack global identification as well. Thus, data routing is a very critical issue in such conditions. Therefore, designing appropriate routing scheme highly depends upon the underlying application requirement. Popular WSN routing schemes are sensor protocols for information via negotiation (SPIN), constrained anisotropic diffusion routing (CADR), active query forwarding in sensor networks (ACQUIRE), low-energy adaptive clustering hierarchy (LEACH), power-efficient gathering in sensor information systems (PEGASIS), and threshold-sensitive energy-efficient sensor network protocol (TEEN) [3–5, 8].

1.5.5 Security

Sensor networks are vulnerable to several key attacks. Most popular are eavesdropping (adversary manages to listen data and communication), denial-of-service attacks (a particular node denies to execute the network tasks), Sybil attack (malicious nodes manage to get multiple identities to disrupt routing, resource allocation, and data aggregation), physical attacks (adversary manages to sensor node tampering), and traffic analysis attacks (adversary manages to reconstruct network topologies) [7, 18, 35]. Therefore, network security is a very essential aspect in the WSN, especially if it deployed in enemy prone or secret environment. Continuous research is going on to propose appropriate defenses to protect the sensor networks against attacks. Speaking in more technical words, the security in the WSN refers to ensure three important data centric aspects:

1. Data confidentiality: It means an adversary must not be able to steal and interpret data.
2. Data integrity: An adversary must not be able to alter or damage data.
3. Data availability: An adversary must not be able to disturb data communication link between source nodes and sink node of the WSN.

1.5.6 Production Costs

As we know the WSN generally consists of several hundreds or even thousands sensor nodes [3, 4, 8]. Therefore, the cost of a single node is crucial to decide the overall cost of the WSN. If deploying the WSN is costlier than deploying traditional sensors, then the WSN is not at all cost justified. Therefore, the cost of each sensor node must be as low as possible for the sensor network to be feasible. Now a day, due to advancement in Bluetooth technology, the cost of a sensor node is around only 1–2$.

1.5.7 Fidelity and Scalability

Scalability broadly refers to how well all the operational specifications of a sensor network are satisfied with a desired fidelity, as the number of nodes grows without bound [3, 4, 8]. Based on the operating environment and the phenomenon to be observed, fidelity can cover various performance parameters, such as spatial and temporal resolution, misidentification probability, consistency in data transmission, latency of event detection, and event detection accuracy. Depending on the measure of fidelity, scalability can be formulated in terms of reliability, network capacity, energy consumption, resource exhaustion, or any other operational parameter as the number of nodes increases. Thus, there exists high level of trade-off between scalability and fidelity. Therefore, one has to decide scalability and fidelity for the designed sensor network, depending upon the application requirement.

1.6 Existing WSN Platforms

History of design and deployment of the WSN dates back to the World War II [3, 4, 8]. A platform of acoustic sensors was developed by the USA to detect and track Soviet submarines for sound surveillance. It is currently used by the National Oceanographic and Atmospheric Administration (NOAA) for detecting and monitoring events, such as seismic and animal activity in the ocean. In 1980, the research on the WSN-entitled distributed sensor networks (DSN) was carried out at DARPA (Defense Advanced Research Projects Agency). The network consisted of many spatially distributed, low cost, autonomous sensing nodes that collaborate among each other for data routing. A number of such research attempts on the design and development of the WSN have been reported in the history. At present there is no such common WSN platform to be used for a specific application. The platform of Berkeley motes and their variants have wider user and developer communities. It is quite less expensive to build our own WSN platform for intended application in mind than to buy commercially available platforms. Therefore, a popular trend to design and produce own WSN setup has been established for the last two decades among many researchers, R&D labs, and commercial companies prefer. Some of these research attempts and related projects are explained in the following sections:

1.6.1 Wins

The University of California in association with the Rockwell Science Center developed wireless integrated network sensors (WINS) project, which was later on commercialized with the Sensoria Corporation (San Diego, California) in 1998 [3, 4, 8]. This project covered almost all the aspects in the WSN design right from MEMS sensor and transceiver integration at the circuit level, network protocol design, and signal processing architectures to the fundamentals of its sensing and detection theory. The project concluded that WINS would provide distributed networking and Internet accessibility to sensor nodes, task controls, and adding embedded processors with the node.

1.6.2 Eyes

The Infineon developed energy-efficient sensor networks (EYES). This project was funded by the European Union (EU) to design and develop the technology and architecture of wireless sensors that can be networked with large number of mobile nodes [3, 4, 8]. The project eyed at supporting devices such as PDAs, laptops, and even mobile phones. The developed sensor nodes are equipped with a TI's MSP430 processor, SAW filter, radio device TDA 5250, and transmission power control. Each sensor node has a USB port for interfacing to a PC. These sensor nodes also have provision to add extra sensors as well as actuators, depending upon the application demand.

1.6.3 Pico-Radio

In 1999, the Pico-Radio project started at the University of California to support the development of low-cost, low-energy sensor nodes with ad hoc capability. The proposed for the Pico-Radio network has physical layer with direct sequence spread spectrum and the MAC protocol with the application of carrier sense multiple access (CSMA) and spread spectrum techniques [4–7]. The important findings of this project are as follows: (1) The node can randomly select a channel and monitor the network activity. (2) If the channel is currently engaged, the node can search for another channel from the list of the remaining available channels. Once an idle channel is detected, the scanning is stopped. (3) In case the idle channel is not found, the node would back off and set a random timeout timer for each channel. (4) It can then use the channel which has first expired timer. Then, the timers for the other channels are cleared off.

1.6.4 Mica Mote Family

The sensor nodes of Mica mote family are developed at the University of California, Berkeley. This project started in partial collaboration with Intel in the late 1990s. These sensor nodes are commonly referred as Mica motes, with different variants such as Mica, MicaZ, Mica2, and Mica2Dot, which are commercially sold via the Crossbow company [3, 4, 6, 7]. The OS in these products is TinyOS. The TinyOS is coded in the nesC language with a component-based protocol. The Mica motes use a processor from Atmel family (usually ATmega128L 8-bit processor running at 7 MHz) and a radio modem from RFM (usually it is TR 1000). In Mica motes sensors are interfaced to the controller using I2C or SPI protocols. Power to Mica motes is provided via two AA batteries of current capacity of 2000 mAh. The Chipcon transceivers are generally employed in Mica motes. For instance, in Mica2 mote has Chipcon CC1000 transceiver, which operates on the 868/915 MHz band with data rate of 38.4 kbps. In MicaZ the Chipcon CC2420 transceiver operating in unlicensed 2.4 GHz band with data rate of 250 kbps is used. It uses offset quaternary phase-shift keying (O-QPSK) as a modulation technique.

1.7 Applications of WSN

Due to the continuous development in the WSN technology, the assets of national importance such as aircrafts, ships, and even buildings can detect structural faults on time (this application area is popularly known as structural health monitoring) [7, 26, 36–38]. The WSN also has paved a way to design and develop systems that provide useful prior alerts before earthquake and tsunami. The WSN also has extensive applications in the battlefield for surveillance and reconnaissance. The WSNs can be used in critical applications such as earthquake, tsunami, battlefield, and flood and also in enemy intrusion detection, target tracking, forest fire detection, industry monitoring, structural monitoring, and environmental and biological monitoring. Although it covers a broad range of diverse application areas, few of them are described below:

1.7.1 Military Applications

The sensor network research was originally motivated from the military needs [3, 4, 6, 8]. The demand in military-based applications includes energy conservation, rapid deployment of assets, as well as robust sensing along the rivers and in hostile terrains. The typical military applications are listed below:

- Monitoring and tracking enemy forces and monitoring friendly forces
- Monitoring equipment and inventory
- Reconnaissance
- Surveillance of war area
- Assessment during war damage
- Nuclear, biological, and chemical attack detection
- National border monitoring

1.7.2 Environment Monitoring Applications

The WSN has been proved to be the ideal choice for many environment monitoring applications due to its capability of unattended operation. The typical environment-related applications are listed below:

- Weather sensing and monitoring stations
- Forest fire detection
- Habitat monitoring
- Monitoring pollution level of water, land, and air
- Flood detection
- Precision agriculture
- Endangered species population measurement
- Tracking migrations of bird and endangered wild animals
- Soil erosion detection

1.7.3 Health Applications

Many times humanly monitoring of patients or medical equipment during complex surgery in big hospitals is impossible [39–41]. The wireless sensors in such situation can assist the doctors and hospital administration to execute various tasks accurately and appropriately. The typical health systems-related applications wherein the WSN is involved are listed below:

- Physiological data monitoring remotely
- Locating and tracking of patients and doctors inside a hospital
- Administrating drug remotely

- Assistance to elderly people

1.7.4 Home Applications

The improvement in the quality of life by creating secure and intelligent living spaces for humans is the underlying idea behind smart homes [39, 42, 43]. The WSN finds a huge potential for applications in the area of smart homes. The typical home automation-related applications are listed below:

- Home automation
- Instrumented environment
- Automated meter reading
- Tracking system for child and elderly people

1.7.5 Other Commercial Applications

Sensor networks are also proved to be highly useful in some of the commercial applications of national importance [39, 42, 43]. In commercially important applications, the WSN can not only provide reliable measurements using which localization of important entities can be done efficiently.

- Monitoring nation's critical resources such as power industrial plants, tunnels, communication grids, and parks
- Ambient temperature control in office and industrial buildings
- Inventory management and control
- Landslide detection systems
- Vehicle tracking and detection
- Traffic flow surveillance on highways
- Air traffic control stations

References

1. Y. Zhang, L. Sun, H. Song, X. Cao, Ubiquitous WSN for healthcare: Recent advances and future prospects. IEEE Internet Things J. (2014). https://doi.org/10.1109/JIOT.2014.2329462
2. S. Gezici et al., Localization via ultra-wideband radios: A look at positioning aspects of future sensor networks. IEEE Signal Process. Mag. (2005). https://doi.org/10.1109/MSP.2005.1458289
3. I. F. Akyildiz, T. Melodia, K. R. Chowdhury, A survey on wireless multimedia sensor networks. Comput. Netw. **51**(4) (2007). https://doi.org/10.1016/j.comnet.2006.10.002

4. I. Khemapech, I. Duncan, A Miller, A survey of wireless sensor networks technology, in *6th Annual Postgraduate Symposium on the Convergence of Telecommunications, Networking and Broadcasting*, vol. 6 (2005)
5. W. Dargie, C. Poellabauer, *Fundamentals of Wireless Sensor Networks: Theory and Practice* (Wiley, New York, 2011)
6. G. Xu, W. Shen, X. Wang, Applications of wireless sensor networks in marine environment monitoring: a survey. Sensors (Switzerland) **14**(9) (2014). https://doi.org/10.3390/s140916932
7. P. Kumar, H. J. Lee, Security issues in healthcare applications using wireless medical sensor networks: a survey. Sensors **12**(1) (2012). https://doi.org/10.3390/s120100055
8. B. Rashid, M. H. Rehmani, Applications of wireless sensor networks for urban areas: a survey. J. Netw. Comput. Appl. **60** (2016). https://doi.org/10.1016/j.jnca.2015.09.008
9. S. R. Jondhale, R. S. Deshpande, GRNN and KF framework based real time target tracking using PSOC BLE and smartphone. Ad Hoc Netw. (2019). https://doi.org/10.1016/j.adhoc.2018.09.017
10. S. R. Jondhale, R. S. Deshpande, Kalman filtering framework-based real time target tracking in wireless sensor networks using generalized regression neural networks. IEEE Sensors J. (2019). https://doi.org/10.1109/JSEN.2018.2873357
11. S. Jondhale, R. Deshpande, Self recurrent neural network based target tracking in wireless sensor network using state observer. Int. J. Sensors Wirel. Commun. Control (2018). https://doi.org/10.2174/2210327908666181029103202
12. S. R. Jondhale, R. S. Deshpande, Modified Kalman filtering framework based real time target tracking against environmental dynamicity in wireless sensor networks. Ad Hoc Sens. Wirel. Netw. **40**(1–2), 119–143 (2018)
13. M. Zhou, Q. Zhang, Z. Tian, F. Qiu, Q. Wu, Integrated location fingerprinting and physical neighborhood for WLAN probabilistic localization, in *Fifth International Conference on Computing, Communications and Networking Technologies (ICCCNT)* (2014). https://doi.org/10.1109/ICCCNT.2014.6963028
14. R. S. Campos, L. Lovisolo, M. L. R. De Campos, Wi-Fi multi-floor indoor positioning considering architectural aspects and controlled computational complexity. Expert Syst. Appl. (2014). https://doi.org/10.1016/j.eswa.2014.04.011
15. A. Payal, C. S. Rai, B. V. R. Reddy, Artificial neural networks for developing localization framework in wireless sensor networks, in *2014 International Conference on Data Mining and Intelligent Computing (ICDMIC)* (2014). https://doi.org/10.1109/ICDMIC.2014.6954228
16. M. Anand, T. Sasikala, Efficient energy optimization in mobile ad hoc network (MANET) using better-quality AODV protocol. Cluster Comput. **22** (2019). https://doi.org/10.1007/s10586-018-1721-2
17. C. Feng, W. S. A. Au, S. Valaee, Z. Tan, Received-signal-strength-based indoor positioning using compressive sensing. IEEE Trans. Mob. Comput. (2012). https://doi.org/10.1109/TMC.2011.216
18. S. R. Jondhale, R. S. Deshpande, S. M. Walke, A. S. Jondhale, Issues and challenges in RSSI based target localization and tracking in wireless sensor networks, in *2016 International Conference on Automatic Control and Dynamic Optimization Techniques (ICACDOT)* (2017). https://doi.org/10.1109/ICACDOT.2016.7877655
19. S. R. Jondhale, R. S. Deshpande, Tracking target with constant acceleration motion using Kalman Filtering, in *2018 International Conference On Advances in Communication and Computing Technology (ICACCT)* (2018). https://doi.org/10.1109/ICACCT.2018.8529628
20. N. Patwari, J. N. Ash, S. Kyperountas, A. O. Hero, R. L. Moses, N. S. Correal, Locating the nodes: Cooperative localization in wireless sensor networks. IEEE Signal Process. Mag. (2005). https://doi.org/10.1109/MSP.2005.1458287
21. S. Kumar and S. Lee, Localization with RSSI values for wireless sensor networks: an artificial neural network approach. Int. J. Comput. Netw. Commun. (2014). https://doi.org/10.3390/ecsa-1-d007

22. Z. Chen, Q. Zhu, and Y. C. Soh, Smartphone inertial sensor-based indoor localization and tracking with iBeacon corrections. IEEE Trans. Ind. Inf. (2016). https://doi.org/10.1109/TII.2016.2579265
23. L. Mihaylova, D. Angelova, D. R. Bull, N. Canagarajah, Localization of mobile nodes in wireless networks with correlated in time measurement noise. IEEE Trans. Mob. Comput. (2011). https://doi.org/10.1109/TMC.2010.132
24. A. El-Rabbany, *Introduction to GPS: the global position system* (Artech House, London, 2006)
25. P. A. Zandbergen, S. J. Barbeau, Positional accuracy of assisted GPS data from high-sensitivity GPS-enabled mobile phones. J. Navig. (2011). https://doi.org/10.1017/S0373463311000051
26. M. B. Higgins, Heighting with GPS: possibilities and limitations, in *Comm. 5 Int. Fed. Surv.* (1999)
27. Z. Bin Tariq, D. M. Cheema, M. Z. Kamran, I. H. Naqvi, Non-GPS positioning systems. ACM Comput. Surv. (2017). https://doi.org/10.1145/3098207
28. F. Viani, M. Bertolli, M. Salucci, A. Polo, Low-cost wireless monitoring and decision support for water saving in agriculture. IEEE Sensors J (2017). https://doi.org/10.1109/JSEN.2017.2705043
29. R. Faragher, R. Harle, Location fingerprinting with bluetooth low energy beacons. IEEE J. Sel. Areas Commun. (2015). https://doi.org/10.1109/JSAC.2015.2430281
30. M. H. Anisi, G. Abdul-Salaam, A. H. Abdullah, A survey of wireless sensor network approaches and their energy consumption for monitoring farm fields in precision agriculture. Precis. Agric. (2015). https://doi.org/10.1007/s11119-014-9371-8
31. P. Abouzar, D. G. Michelson, M. Hamdi, RSSI-based distributed self-localization for wireless sensor networks used in precision agriculture. IEEE Trans. Wirel. Commun. (2016), https://doi.org/10.1109/TWC.2016.2586844
32. J. Yick, B. Mukherjee, D. Ghosal, Wireless sensor network survey. Comput. Netw. (2008). https://doi.org/10.1016/j.comnet.2008.04.002
33. R. Silva, J. Sa Silva, F. Boavida, Mobility in wireless sensor networks - survey and proposal. Comput. Commun. (2014). https://doi.org/10.1016/j.comcom.2014.05.008
34. V. C. Paterna, A. C. Augé, J. P. Aspas, M. A. P. Bullones, A bluetooth low energy indoor positioning system with channel diversity, weighted trilateration and kalman filtering. Sensors (Switzerland) (2017). https://doi.org/10.3390/s17122927
35. Y.W. Prakash, V. Biradar, S. Vincent, M. Martin, A. Jadhav, *Smart bluetooth low energy security system* (2018). https://doi.org/10.1109/WiSPNET.2017.8300139
36. M. S. Pan, Y. C. Tseng, ZigBee wireless sensor networks and their applications. Sens. Netw. Config. Fundam. Stand. Platforms Appl. (2007). https://doi.org/10.1007/3-540-37366-7_16
37. M. R. Mohd Kassim, I. Mat, A. N. Harun, Wireless sensor network in precision agriculture application, in *2014 International Conference on Computer, Information and Telecommunication Systems (CITS)* (2014). https://doi.org/10.1109/CITS.2014.6878963
38. A. Minaie, Application of wireless sensor networks in health care system application of wireless sensor networks in health care system, in *ASEE Annual Conference and Exposition* (2013)
39. R. J. F. Rossetti, Internet of Things (IoT) and smart cities, in *IEEE Readings Smart Cities* (2015)
40. A. Zanella, Best practice in RSS measurements and ranging. IEEE Commun. Surv. Tutorials (2016). https://doi.org/10.1109/COMST.2016.2553452
41. B. Latré, B. Braem, I. Moerman, C. Blondia, P. Demeester, A survey on wireless body area networks. Wirel. Netw. (2011). https://doi.org/10.1007/s11276-010-0252-4
42. F. Viani, P. Rocca, G. Oliveri, D. Trinchero, A. Massa, Localization, tracking, and imaging of targets in wireless sensor networks: an invited review. Radio Sci. (2011). https://doi.org/10.1029/2010RS004561
43. D. M. Han, J. H. Lim, Smart home energy management system using IEEE 802.15.4 and zigbee. IEEE Trans. Consum. Electron. (2010). https://doi.org/10.1109/TCE.2010.5606276

Chapter 2
Target Localization and Tracking Using WSN

2.1 Introduction to WSN-Based L&T

Indoor L&T of target is useful for many applications in several sectors, such as the manufacturing, sports, healthcare, and construction [1–6]. For instance, in the healthcare sector, locating and tracking the locations of objects can be very crucial whenever and wherever there is high emergency to respond to. For instance, in hospitals many employees need to share the same hospital assets during work. These items many times are moved from their regular location and not returned to the original location after the work is finished. In the manufacturing sector, the knowledge of location of the finished products and other items in a warehouse can help in asset management by keeping a track on the inventory and lowering the searching time to find them. The person unfamiliar with the given built environment (e.g., a large building) can be provided with location map of the area in order to search out the routes toward intended destination. Critical information objects (e.g., objects in museums or computer hard-drives) or high-valued assets can also be exposed to possible thefts. In other words, location knowledge can provide theft detection as well as prevention by giving useful alerts about whenever they are shifted outside from the predefined boundaries by some unauthorized persons.

Target L&T is one of fundamental applications of the WSN, wherein the main objective is to detect and locate the mobile target and also to keep a track on its movement path (trajectory) continuously with the help of field measurements from sensor nodes [7–10]. This is termed a single target L&T problem; although if the problem involves L&T of multiple mobile targets with the help of WSN-based setup, then it is termed as a multi-target L&T problem. The low maintenance cost, simple and random deployment procedure, ad hoc nature, and the possibility of unattended operation make WSN a vital option for various indoor L&T applications. The WSN can easily locate and track the trajectory of the moving target by

S. R. Jondhale et al., *Received Signal Strength Based Target Localization and Tracking Using Wireless Sensor Networks*, EAI/Springer Innovations in Communication and Computing, https://doi.org/10.1007/978-3-030-74061-0_2

simply exploiting field measurements once deployed randomly. In these problems, the sensor nodes are deployed at random or predefined locations in the sensing environment.

Consider a general target L&T scenario using WSN as shown in Fig. 2.1, wherein the target is moving inside the WSN monitored area along a predefined or unknown path. The sensor nodes in the WSN based L&T system are designated as detecting nodes (the nodes which are in the vicinity of the mobile target and able to detect the target), vigilant nodes (the nodes which are likely to detect the target in the future), and inactive nodes (which are not at all utilized in the L&T process). The mobile target can be any object, such as asset, an animal, an intruder, a vehicle, or a person [7–10]. Figure 2.2 shows the basic procedural steps executed in target L&T mechanism. It consists of detection of a target during its motion in the WSN monitored area, localization to locate the mobile target, and tracking to trace the route of mobile target.

The WSN-based L&T may also be classified as single target vs multiple target L&T, active vs passive L&T, indoor vs outdoor L&T, and two-dimensional (2-D) vs three-dimensional (3-D) L&T [7, 11–14]. If the target cooperates in localization, then it is termed as active L&T; otherwise, it is called as passive L&T. In the former case, the target is with a sensor node, and the rest of the WSN nodes can detect and locate the target. In the latter case, the target is "device-free," wherein the target is not equipped with a WSN node. This book is intended to discuss the design and development of L&T algorithms to efficiently track a single mobile target in an indoor environmental setup by exploiting field measurements.

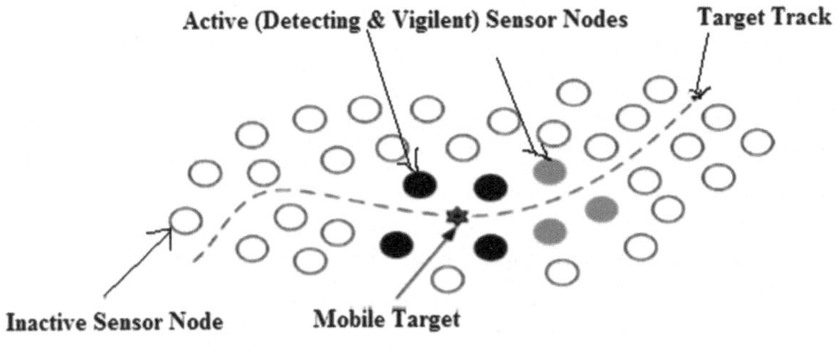

Fig. 2.1 General target L&T scenario using WSN

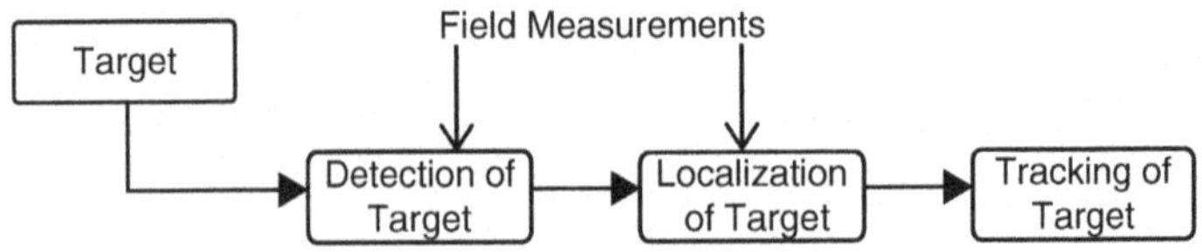

Fig. 2.2 General mechanism of target L&T

The dramatic technological revolution in smartphones, wearable wireless devices, and WSN in the last decade has come up with a wide variety of useful applications, including indoor L&T applications [7, 11–14]. Indoor L&T is the process of achieving user location, which can be utilized in a wide range of applications in health sector, disaster management, smart home, and surveillance. It is also proved to be beneficial in many important areas, such as smart cities, smart structures, and smart grids. In the context of the WSN-based L&T for the indoor setup, there are two types of sensor nodes, namely, anchor nodes (also called as reference nodes) and non-anchor nodes [7]. Generally, the anchor nodes are deployed at known locations, whereas locations of non-anchors nodes are unknown. The moving target is assumed to carry one non-anchor node. The target locations during its movement are estimated with the help of anchor nodes through internode communications. However, environmental issues, such as signal fading, multipath propagation, and non-line of sight (NLOS), pose the major challenges in achieving high tracking accuracy. The WSN-based tracking systems must also be robust enough to deal with abrupt variations in target velocity as well as variation in target mobility pattern. Therefore, the researchers from academia and industry need to propose efficient L&T algorithms with reference to the challenges mentioned above.

2.1.1 Typical L&T Scenario in Wireless Sensor Networks

The target L&T in an indoor environment using WSN enables a wide variety of applications [1–4]. As discussed earlier, the mobile target can be any object, such as an asset, an animal, an intruder, a vehicle, or a person. Sometimes the mobile target moves along a predefined path, and sometimes the target path is unknown. A typical scenario of target L&T using WSN is shown in Fig. 2.3. The target state at any time instance k can be given by the state vector $X_k = (x_k, y_k, \dot{x}_k, \dot{y}_k)$, wherein x_k and y_k are the target locations and $\dot{x}_k$ and $\dot{y}_k$ are the target velocities in x and y directions, respectively. One may augment acceleration parameters x_k and y_k along x and y directions, respectively, in the above the state vector. During the target motion in the WSN, the state vector changes. The objective of the deployed WSN is to estimate continuously the state vector using field measurements from the environment with the help of a suitable L&T algorithm [11, 15–18]. Thus, for the case of mere target localization, it is one time estimation problem, whereas for the case of target tracking, it is a sequential state estimation problem. That means the algorithms that are used for target tracking problem are the same as that for the target localization problem. At the end of a state vector estimation, we are interested to know about how the L&T algorithm performed in the context of target L&T for the considered system design and assumptions. The performance evaluation parameters that are generally used for target L&T are localization error, RMSE, or both. The lower the values of these performance evaluation parameters, the higher will be the target L&T accuracy.

As discussed in the last paragraph, the target tracking being a sequential localization problem, it needs the location estimation algorithm of recursive nature [16–19].

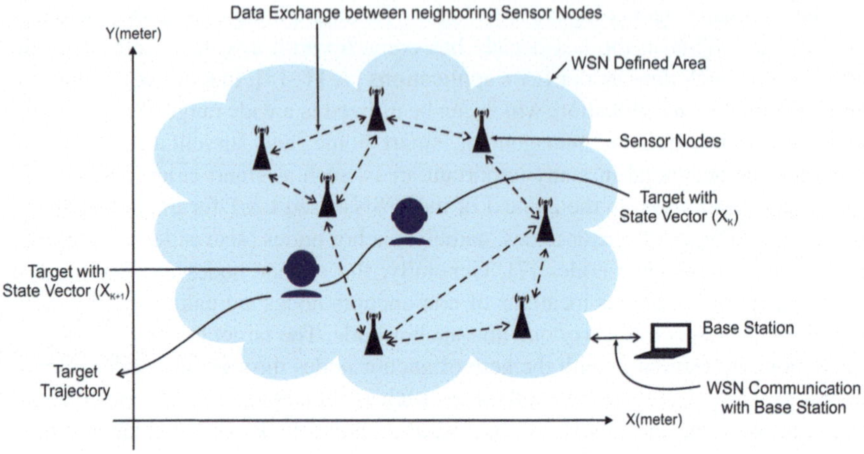

Fig. 2.3 Typical target tracking scenario using WSN

One may call this recursive location estimation algorithm as the target tracking algorithm. Several factors that impact the performance of the target tracking algorithm are the following: the type of environment (indoor/outdoor), type of field measurement involved, density of obstacles in the environment, algorithmic design, and density of anchor and non-anchor nodes. Apart from these system design issues, the field measurement also faces the problem of signal propagation issues, such as signal fading, reflections, NLOS conditions, and multipath propagation. Therefore, to develop a robust and high precision target L&T system for an indoor environment is a highly challenging task. Due to such environmental dynamicity, the existing target L&T systems suffer with a low L&T accuracy (i.e., if localization error is higher than 1 m, then it can be considered as low L&T accuracy). In addition to issues of system design and environmental dynamicity, some other issues, such as abrupt changes in target velocity, during motion and availability of less field measurements can also deteriorate the performance of the L&T algorithm further. Therefore, research has been continuously going on to design and develop robust and accurate target L&T systems, which can offer higher target localization accuracy (i.e., localization error lower than 1 m), real-time performance, and lower computational simplicity.

2.1.2 Classification of Target L&T Techniques

As discussed several times previously, the WSN utilizes the field measurements to locate the mobile target. Based on the involvement of distance of the target from the anchor nodes in the computation or estimation of the unknown locations of the mobile target during its motion, the L&T algorithms can be divided into two major classes: range-based L&T and range-free L&T as shown in Fig. 2.4 [9, 11, 20]. If the target L&T algorithm depends upon the distance (range) between target and

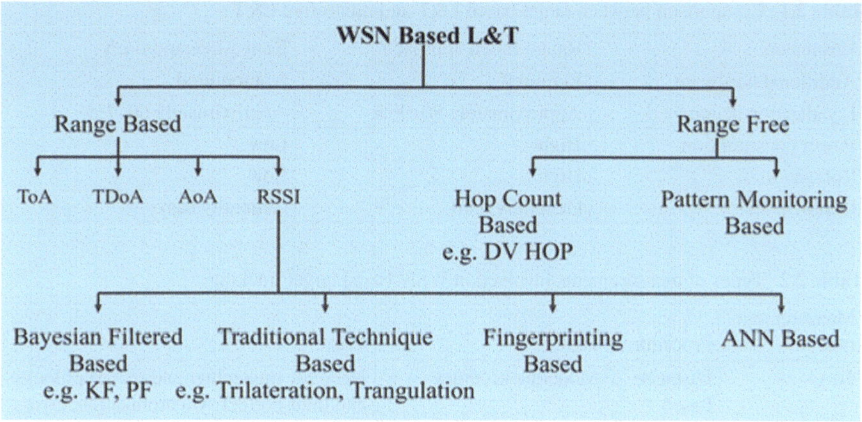

Fig. 2.4 Classification of WSN-based L&T

anchor nodes during estimation, then it is termed as range-based algorithms; otherwise, it is termed as range-free algorithms. Unlike range-based approach, in the range-free approach the connectivity of sensor nodes is utilized to locate the moving target rather than the distance between the target and anchor node. The target L&T accuracy of range-based algorithms is generally high as compared to its counterpart range-free algorithms. However, looking from hardware perspectives, the range-free algorithms require additional hardware when compared with range-free algorithms. The overall comparison between range-free and range-based approaches is presented in Table 2.1.

The range-based approach utilizes field measurements, such as the time of arrival (ToA), angle of arrival (AoA), received signal strength indicator (RSSI), and time difference of arrival (TDoA) [7, 21]. In the AoA, the angles of arrival of signals between target and anchor nodes are utilized to locate the moving target. Although the AoA technique does not need clock synchronizations between transmitters and receivers, the need of an array of directional antennas is its main limitation. In the ToA-based L&T approach, the signal propagation velocity and the time of arrival of the transmitted signal are exploited to calculate the distances from transmitter to receiver, whereas in the TDoA-based approach, the time difference of arrival of signals coming from the transmitter and receiver is utilized. The major drawback of TDoA and ToA techniques is the need of exact time synchronization between the transmitter and receiver clocks, the susceptibility to NLOS conditions, interferences, and measurement noise. The additional hardware requirements in AoA, ToA, and TDoA range-based techniques make the L&T system expensive and bit complex. Unlike AoA, ToA, and TDoA range-based L&T approaches, in the RSSI-based L&T approach, there is no such requirement of additional hardware for the target L&T. In the RSSI L&T approach, the distance between the target and anchor nodes using a suitable path loss model is utilized to locate the target. The prerequisites for the path loss model are knowledge of the transmitted and received signal powers, transmitting

Table 2.1 Comparison between range-based L&T and range-free L&T

Parameters	Range-based approach	Range-free approach
Additional hardware	Required	Not required
Localization accuracy	Approximately 80–90%	Approximately 60–75%
Power consumption	High	Low
Robustness	High	Low
Deployment	Generally hard	Generally easy

Table 2.2 Types of measurements involved in WSN-based target tracking

Measurement type	Procedure	Pros	Cons
ToA	Distance-based	Moderate accuracy	Need for transmitter and receiver clocks and their perfect synchronization; errors due to NLOS conditions, signal noise, and interferences
TDoA	Distance-based	High accuracy	Need for transmitter and receiver clocks and their perfect synchronization; errors due to NLOS conditions, signal noise, and interferences
AoA	Angle-based	High accuracy	Requirement of directional antenna array
RSSI	Distance-based	No need for additional hardware, low cost, and low power consumption	RSSI measurements are susceptible to environmental dynamicity and moderate accuracy

and receiving antenna gains, and operating frequency. The pros and cons associated with these field measurements are given in detail in Table 2.2.

The range-free techniques are classified as hop count-based technique (e.g., DV Hop) and pattern matching-based technique (e.g., approximate point in triangle (APIT)) [11, 20, 22, 23]. Basically, these both approaches are area-based methods. In DV-Hop-based L&T approach, the unknown location of node (or target) is computed by counting the number of hops the RF signal takes to reach the destination. In APIT-based L&T approach, the information such as whether the node (or target) is within a predefined area or not is utilized. These both approaches do not provide the exact location of the target; instead of that, they provide the area in which the target is. An artificial neural network (ANN) can be used in both range-based and range-free methods. As this book is intended to discuss the fundamentals of the RSSI-based target L&T approach only, the rest of the other approaches are out of the scope of this book. The detailed discussion of the RSSI-based target L&T approach is discussed in detail in the next section.

2.2 RSSI-Based Target L&T Approach

The RSSI is basically the measure of the magnitude of power received at the receiver terminal. The RSSI measurements during RF communication are obtained very easily at the receiver during normal communication [11, 15, 16, 24, 25]. As discussed in the previous section, the L&T system based on the RSSI measurements neither needs an array of directional antennas nor needs synchronization between the receiver and transmitter clocks. Each wireless sensor node is with on-chip RSSI circuit, which can give the values of RSSI measurements. Thus, there is no need for additional hardware in the RSSI-based target L&T approach. Hence, the RSSI metric has been dominantly used in the WSN-based target L&T systems. Compared to other counterparts, few other important advantages associated with the RSSI-based L&T approaches are as follows: simpler procedural aspects and lower power consumption.

Theoretically speaking, the RSSI is a function of distance between the receiver and transmitter and the RF environment, in which the WSN or other wireless system is deployed. Therefore, due to the dependence on the RF channel, the RSSI-based L&T algorithms are generally affected by changes in the environmental setup [11, 15, 16, 24, 25]. In fact, in the RSSI-based approach, the distance between the receiver and the transmitter is computed using the difference between magnitudes of transmitted power and that of received power. This power difference is termed as signal attenuation or path loss. Therefore, the utmost care is to be taken to choose an appropriate path loss model to characterize the given RF channel. Speaking in more technical words, the RSSI is a part of the IEEE 802.11 protocol family. The RSSI values are measured in dBm unit. The RSSI values generally fall between 0 dBm (excellent signal) and −110 dBm (very poor signal) and are negative [26, 27]. In the indoor L&T applications with WSN, the RSSI measurement-based approach is generally used as compared to the rest of the alternatives. In these applications for indoor environmental setup, the aspects that are of prime importance to the success of the underlying application are the following: selections of path loss model, density and locations of non-anchor and anchor nodes, selection of suitable transmission power level algorithmic design, and issues related with signal propagation, such as fading, reflections, NLOS conditions, and multipath propagation [15, 16, 19, 28, 29].

In most of the RSSI-based indoor L&T applications, it is assumed that the target carries one sensor node, which is configured in the transmitter mode, whereas the rest of the sensor nodes of the WSN are configured in the transceiver mode. All the RSSI-based target L&T algorithms discussed in this book from Chap. 4 onward are based on this assumption, although some applications in the literature also assumes the target configured in the receiver mode to collect the measurements from the surrounding sensor nodes. In the first case, the target broadcasts RF signal in the surrounding WSN environment, while the rest of the sensor nodes in the network collect the RSSI measurements of this broadcasted signals. Using the collected RSSI measurements, the distance between the target and the sensor node can be

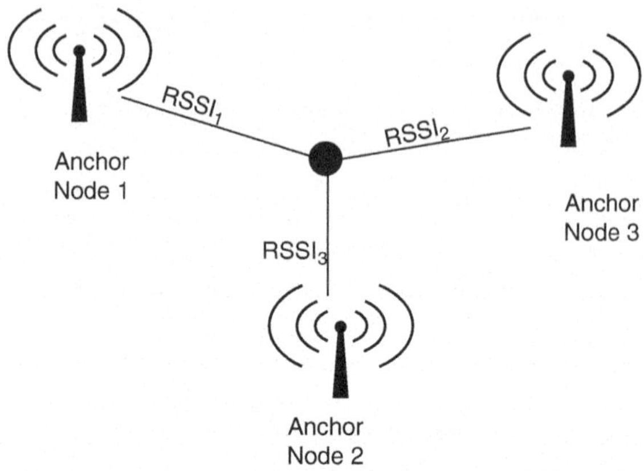

Fig. 2.5 RSSI measurements for target L&T

computed using a suitable signal path loss model. Let's consider a typical scenario showing the use of RSSI measurements to obtain the unknown location of the target as shown in Fig. 2.5. If the target is in the communication range of the three transmitting nodes (anchor nodes), then at the target (which carries a sensor node configured in the receiver mode) three RSSI measurements are received. Then, by using a suitable signal path loss model, one can very easily get three distances of the target from these three anchor nodes. Using these coordinates of the three anchor nodes and three computed distances, the unknown location of the target can be computed very easily. The lower the actual distance between the anchor node and the target node, the higher will be the value of the RSSI measurement and vice versa [11, 30, 31]. The received RSSI measurement is found to have a highly nonlinear relationship with the distance as shown in Fig. 2.6.

The RSSI measurements are generally erroneous due to the issues related with signal propagation, such as attenuation, reflections, fading, NLOS conditions, and multipath propagation [11, 30, 31]. In fact, the RF wave reaches the destination along the different paths of varying length (multipath propagation), and thereby it takes different travel times along these paths. Thus, these components of the same RF signal reach the destination at different times with varying amplitudes. The interaction of these RF components with each other causes multipath fading. That means, these components interfere with each other. These interferences at the receiver can be destructive or constructive. The major reason of multipath propagation and fading is the varying amount of obstacles in the given environment along different paths, and thereby, the RF signal components, along each path, experience varying amount of reflections. The NLOS is the condition wherein the antennas of the transmitting and the receiving nodes are not along a LOS. Therefore, the received RSSI measurements are not reliable, though environment is kept unchanged.

The slight changes in location of experiment can also cause variations in the amount of attenuation, reflections, fading, multipath propagation, and NLOS. Even with the same environmental setup, the same RSSI measurements are not guaranteed. In other words, there exists less possibility of repeatability and regularity in the RSSI measurements. Thus, the RSSI measurements are highly notorious and

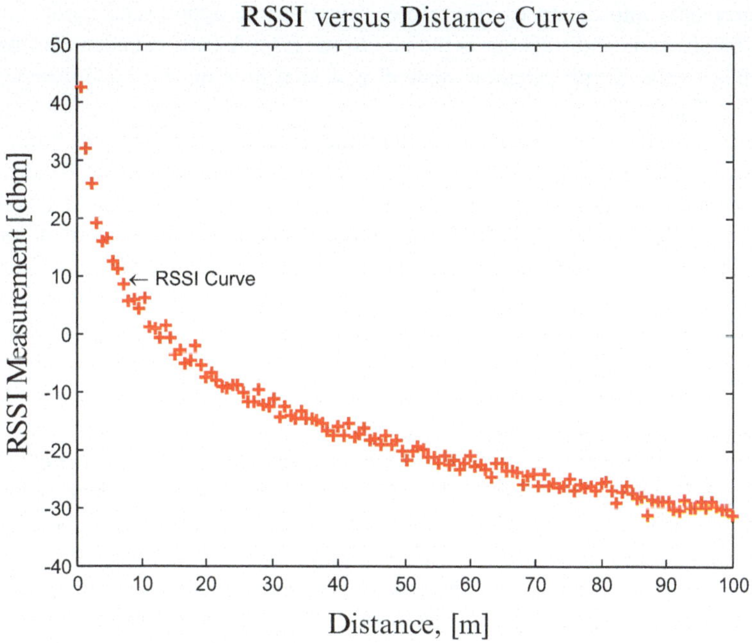

Fig. 2.6 Nonlinear relationship between RSSI and distance

dependent on the environment setup. Due to all these characteristic features and limitations in the RSSI measurements as discussed above, the RSSI-based L&T system is generally associated with low localization accuracy and low stability [11, 30–33]. In order to avoid this problem, some of the precautionary measures reported in the literature are as follows:

- Take the RSSI measurements at several frequency.
- Take the average RSSI measurements over a suitable time period to smooth variations in the RSSI measurements.
- Calibrate WSN transceivers to get a comparable reception sensitivity and emission power.
- Use high-quality antennas.
- Try to minimize changes in the environment setup and signal interference from the surrounding electronic gadgets, rain, and mobile objects.

2.3 Environmental Characterization Through Path Loss Models

As discussed earlier to locate a target using RSSI-based technique, characterization of the given RF channel is a must, and for its characterization, the selection of a suitable path loss model is highly essential. The path loss model translates the RSSI measurements into distances. Therefore, selecting the appropriate model for the target localization and tracking is the key to success. A correct understanding and modeling of the RF propagation channel is a vital prerequisite for improving the target L&T accuracy.

Basically, a path loss model is a set of mathematical expressions, algorithms, and diagrams, which represents the radio characteristics of the considered RF environment in which the target resides [9, 14, 26, 33]. The empirical models of a path loss model are based on the actual RSSI measurements, whereas the theoretical models of a path loss model are based on the fundamental principles of RF communication. Popular RSSI path loss models for RF environment characterization are the following: free space propagation model, log normal shadowing model (LNSM), and two-ray ground model. The modified versions of these basic models have also been reported in the literature. Few researchers in L&T domain have also designed their own path loss models to characterize the given wireless environment. For instance, the authors in [9] have presented the optimal fitting parametric exponential decay model (OFPEDM). The OFPEDM is developed for large-scale wheat field. The author claim that the OFPEDM has less susceptibility to variations in the RF environment and higher distance estimation accuracy. The free space propagation model and two-ray ground model have specific requirements for the underlying application environment, whereas the LNSM model is more general in nature. The LNSM is sometimes also called as log normal shadow fading model. Out of all of these, the LNSM is more suitable in RSSI-based L&T applications for indoor as well as outdoor environmental setup. It presents a number of configurable parameters using which the given RF environment can be artificially simulated. Let's discuss mathematics of all of these path loss models in detail.

2.3.1 Free Space Path Loss Model

The free space path loss model provides the RSSI measurement if the transmitter and the receiver are along a LOS without any obstacle in between [9, 26, 34, 35]. This model is basically based on a well-known Friis transmission formula. It relates the antenna gains, free space path loss, and wavelength to the transmitted and received powers. This equation is one of the fundamental equations in RF communication and antenna theory. In this mathematical equation, if d is the distance between the receiver and the transmitter, then the RSSI measurement at the receiver is denoted as $P_r(d)$. According to this model, the ratio of received power to transmitter power is given as the following:

$$\frac{P_r}{P_t} = \left[\frac{\sqrt{G_t \times G_r}\lambda}{4\pi d}\right]^2 \rightarrow P_r(d) = \frac{P_t G_t G_r \lambda^2}{(4\pi)^2 d^2} \tag{2.1}$$

where P_t and $P_r(d)$ are the transmitted power and the received power, respectively; G_t and G_r are the transmitter antenna gain and the receiver antenna gain, respectively; and λ is the signal wavelength in meters. By rearranging the above equation, one can easily obtain the distance between the transmitter and the receiver.

The Friis equation states that more power is lost at higher frequencies, which is a fundamental result of this equation. In other words, it can be stated that for antennas with some specified gains, the power transfer will be highest at lower frequencies. That means the higher the frequencies, the higher would be the path loss associated. As accurate LOS between the transmitter and the receiver is not always the reality in most of the cases, the estimated RSSI measurements with the help of this model are not reliable, and thus it generally leads to high localization error in the target L&T applications.

2.3.2 Two-Ray Ground Model

The major drawback of free space path loss model is the dependence on the LOS between the receiver and the transmitter [9, 26, 34, 35]. The two-ray ground model does not necessitate the requirement of the LOS. It is basically based on geometry of the given RF environment and pays attention to the direct path as well as the ground reflected path between the receiver and the transmitter (see Fig. 2.7). The estimated RSSI using the two-ray ground model is fairly accurate as compared to that using the free space path loss model. According to the two-ray ground model, the RSSI (received power) is given as [34, 35]:

$$P_r(d) = P_t G_a G_b \frac{h_t^2 h_r^2}{d^4} \tag{2.2}$$

where G_a and G_b are the receiver and the transmitter antenna gain, respectively, and h_r and h_t are the heights of receiver antenna and transmitter antenna, respectively. By rearranging the above equation, one can easily obtain the distance between the transmitter and the receiver.

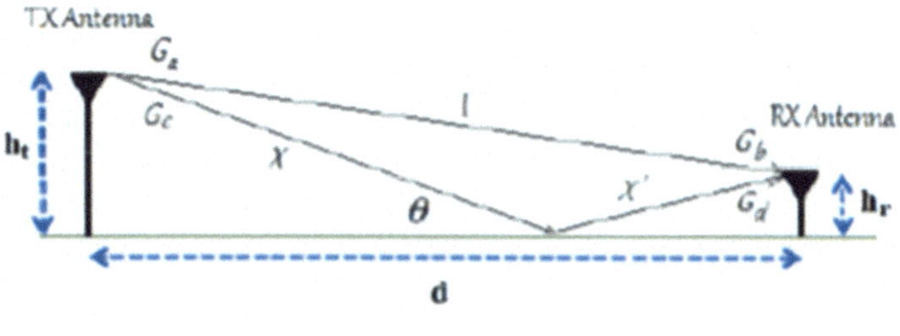

Fig. 2.7 Two-ray ground model

Table 2.3 Path loss exponent (η) for various environments

RF environment		η
Outdoor	Free space	2
	Urban area cellular radio	2.7–3.5
	Shadowed urban area	3–5
Indoor	LOS inside a building	1.6–1.8
	Obstructed in factory	2–3
	Obstructed in building	3–6

2.3.3 *Log Normal Shadow Fading Model (LNSM)*

The LNSM is generic in nature and can be considered as the next version of free space path loss model [34, 35]. It is used to estimate or predict the path loss for a wide variety of RF environments, while the application of the free space path loss model is limited to only a LOS path between the receiver and the transmitter. The path loss *PL*(*d*) at a distance of *d* using LNSM model is given as [16, 19, 24]:

$$PL(d) = P_r(d_0) - 10\ \eta \log\left(d\ /\ d_0\right) + X_\sigma\ , \tag{2.3}$$

where

- $P_r(d_0)$ is RSSI at some reference distance d_0 (generally 1 m).
- X_σ is a normal random variable with the standard deviation of σ (it is a measure of measurement noise due to signal propagation problems arising out of multipath propagation and NLOS).
- η is the path loss exponent (Table 2.3).

The distance *d* between the receiver node and the transmitter node can be computed using Eq. (2.4) as shown below.

$$d = d_0 10^{\left(P_r(d_0)\ -\ PL(d)\ +\ X_\sigma\right)\ /\ 10n} \tag{2.4}$$

2.3.4 *OFPEDM*

As discussed at the start of this section, the signal attenuation model plays an important role in deciding the reliability and quality of ranging (distance computation) accuracy. Hence, some researchers have investigated the signal attenuation model in the actual working environment. The authors in [9] has developed the OFPEDM through the investigation and analysis of localization experiments performed in the wheat field. The general form of the original OFPEDM can be given as follows:

$$PL(d) = X f^Y d^Z \tag{2.5}$$

where

- f is RF signal frequency.
- d is the distance between the receiver and the transmitter.
- X, Y and Z are the constants. These can be computed using regression analysis.

The distance d in Eq. (2.5) can be computed as follows:

$$d = \left(\frac{PL(d)}{Xf^Y} \right)^{1/Z} = \left(\frac{\overline{|RSSI|}}{Xf^Y} \right)^{1/Z} \tag{2.6}$$

$$\overline{|RSSI|} = \frac{1}{N} \sum_{i=1}^{N} RSSI_i \tag{2.7}$$

where $RSSI_i$ represents the RSSI measurement value.

Unlike the LNSM, in the OFPEDM, there is no variable term to account for any measurement noise uncertainty [9]. It is one of the important drawbacks of OFPEDM.

2.4 Technologies for RSSI-Based L&T

This section presents several dominant existing technologies used in indoor L&T services. The important radio communication technologies are Bluetooth, IEEE 802.11, radio frequency identification (RFID), Zigbee, and UWB. Although the camera or vision-based localization systems are also possible, these are beyond the scope of this book.

2.4.1 RFID

The RFID is based on the IEEE 802.15 wireless standard [36–38]. The operating frequency of the RFID is in ultrahigh frequency (UHF), which ranges from 300 to 1000 MHz. However, for RFID applications, two frequency ranges 433 MHz and 860–960 MHz are used. For active tags, 433 MHz frequency is used, whereas for passive tags, 860–960 MHz frequency range is used [39–43]. The RFID refers to a wireless system that uses RF waves at several different frequencies and is consists of two important components: readers and tags. The reader is a device which has one or more antennas. It emits RF waves and receives it back again from the RFID tag. The tag is a device that communicates its identity and other relevant information to nearby readers in the surrounding environment. The tag can be passive or active. The active RFID tags operate on batteries, whereas the passive RFID tags do not need battery and are operated by the reader. The RFID-based L&T has been widely used by research community. This technology utilizes the RSSI measurements for L&T. However, it is observed that indoor environment layout and other setting have direct effects on the reliability of RSSI measurements, which thereafter have impact on the localization accuracy with the underlying applications [11, 19, 24, 26, 27].

The advantage of RFID tag is that it can store several pages of information with each page numbered in order. The reader is generally portable. Sometimes, they are mounted on a post. It can be used on various applications, such as inventory control, out-of-bed detection, equipment tracking, fall detection, and monitoring patients.

2.4.2 Wi-Fi

The Wi-Fi is based on the IEEE 802.11 standard [44–46]. It works in the industrial, scientific, and medical (ISM) band. It basically supports the Internet-based networking capabilities to different wireless devices in public, private, and commercial environments. Its reception range has now been increased from 100 m to about 1 km at the moment. Most of today's laptops, smartphones, as well as various other portable devices are equipped with Wi-Fi facility. Even the existing Wi-Fi ports can also be utilized as anchor nodes in L&T process. Thus, there is no need of additional infrastructure for L&T. That's why Wi-Fi became an ideal candidate for indoor L&T [8, 44–46]. It utilizes the RSSI measurements for L&T applications. But unfortunately, the existing Wi-Fi ports are generally used for data communication rather than L&T purposes; thus, efficient and cost-effective L&T algorithms are needed to be designed to improve localization accuracy. The common Wi-Fi based L&T approaches include the use of methods such as trilateration or triangulation or fingerprinting or their combination.

2.4.3 Bluetooth

Bluetooth is based on the popular IEEE 802.15.1 standard [33, 47–49]. It consists of the physical layer and MAC layer for connecting fixed or mobile wireless devices in a limited area. The modern version of Bluetooth is Bluetooth Low Energy (BLE) (also called as Bluetooth Smart). The BLE is at present a very popular technology for L&T due to low cost, high communication range, and low power consumption. The BLE as against old Bluetooth version is highly energy efficient and can give a—data rate of 24 Mbps and a coverage range of 60–90 m. It can be used with AoA, RSSI, and ToF. However, many existing BLE-based L&T systems utilize the RSSI measurements due to various advantages associated with RSSI as discussed earlier [8, 24, 47, 50, 51]. The two popular recent BLE-based protocols are Eddystone (by Google Inc.) and iBeacon (by Apple Inc.). The typical beacon-based L&T architecture is shown in Fig. 2.8.

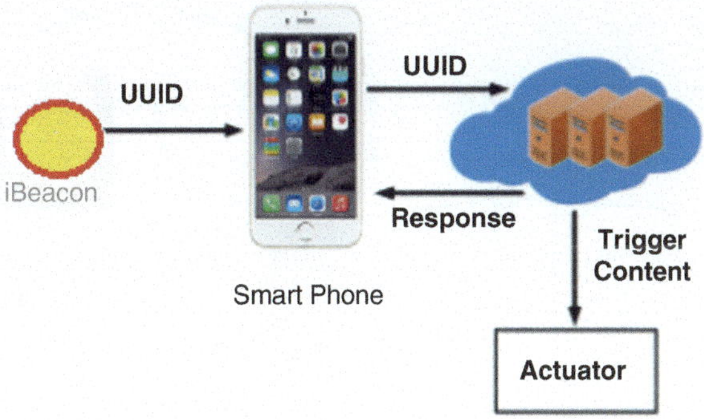

Fig. 2.8 Architecture of iBeacon-based system

The user device contacts server or cloud to recognize the action of the receiver beacon once a message is received from iBeacon. The action can be to open a door, to send a discount coupon to user device, to display some text on monitor, etc.

2.4.4 *Zigbee*

The Zigbee protocol is low power technology and is based on the IEEE 802.15.4 standard [52, 53]. It is basically designed and developed for wireless network monitoring. It works in unlicensed 2.4 GHz frequency band. It has drawn a significant attention in the last decade because of robust communication nature, self-organization and configuration ability, and self-healing capability [14, 52, 53]. It is prone to the external effects and is heavily dependent on the context. These drawbacks of Zigbee can cause serious threats to the underlying L&T applications.

2.5 Traditional Techniques for Target Localization

The popular traditional techniques used for L&T are lateration and angulation [50, 54–56]. In the lateration-based technique, the distances between sensor nodes are utilized for L&T, whereas in angulation technique, the angles of arrival between sensor nodes are utilized for L&T. However, in practice, the computed angles and distances are generally inaccurate due to the variations in system dynamics and noise in the RSSI measurement. The details of these two techniques are presented below in brief.

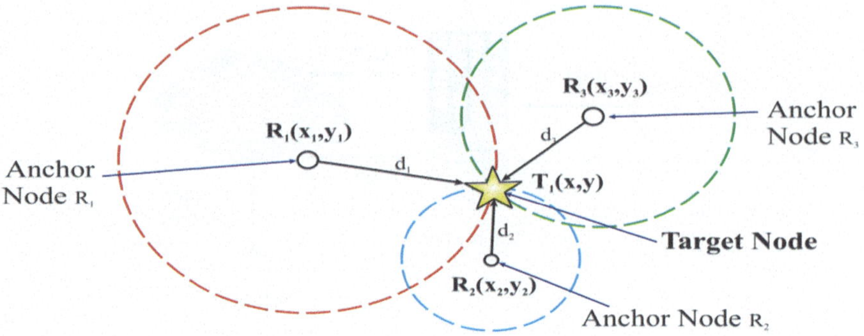

Fig. 2.9 Trilateration-based location estimation

2.5.1 *Trilateration*

Trilateration is basically the process of obtaining the position of a target using its distances (computed using a suitable path loss model) from three anchor nodes. Three circles are formed based on these computed distances and intersection of them is used to locate the target node in space [50, 54–56]. In Fig. 2.9, three anchor nodes R_1, R_2 and R_3 are randomly deployed and a target node is moving around them. Let's consider at particular time instance during motion, the target 2-D location is (x, y) and is at distances d_1, d_2 and d_3 from R_1, R_2 and R_3 respectively. This unknown target location can be computed using the coordinates of anchor nodes and the distances according to Eqs. (2.8)–(2.11).

$$\begin{aligned} d_1^2 &= (x_1 - x)^2 + (y_1 - y)^2 \\ d_2^2 &= (x_2 - x)^2 + (y_2 - y)^2 \\ d_3^2 &= (x_3 - x)^2 + (y_3 - y)^2 \end{aligned} \tag{2.8}$$

If we rearrange Eqs. (2.8) and solve it for x and y, we get:

$$x = \frac{AY_{32} + BY_{13} + CY_{21}}{2(x_1Y_{32} + x_2Y_{13} + x_3Y_{21})}, \quad y = \frac{AX_{32} + BX_{13} + CX_{21}}{2(y_1X_{32} + y_2X_{13} + y_3X_{21})} \tag{2.9}$$

where

$$A = x_1^2 + y_1^2 - d_1^2, \quad B = x_2^2 + y_2^2 - d_2^2, \quad C = x_3^2 + y_3^2 - d_3^2 \tag{2.10}$$

$$\begin{aligned} X_{32} &= (x_3 - x_2), \quad X_{13} = (x_1 - x_3), \quad X_{21} = (x_2 - x_1), \\ Y_{32} &= (y_3 - y_2), \quad Y_{13} = (y_1 - y_3), \quad Y_{21} = (y_2 - y_1). \end{aligned} \tag{2.11}$$

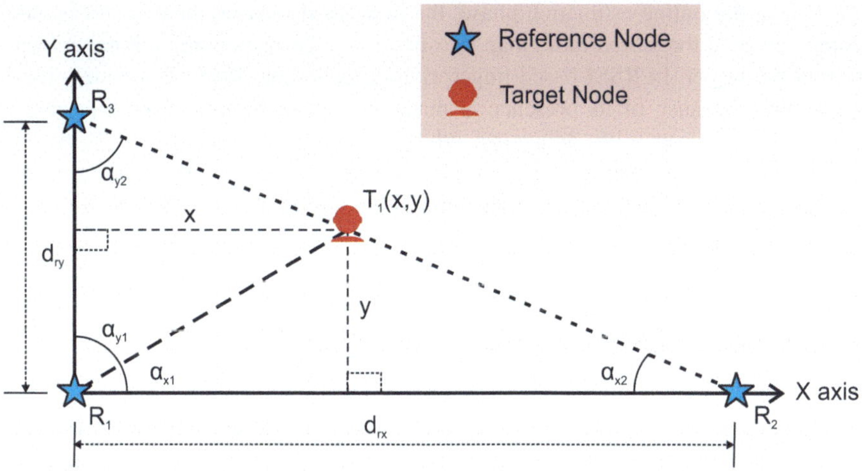

Fig. 2.10 Triangulation-based location estimation

2.5.2 *Triangulation*

Triangulation-based localization is a trigonometry-based approach for L&T. In this method, the location estimations are computed using a distance between deployed anchor nodes and two angles. In this approach, two anchor nodes are needed to be deployed on a horizontal base for x-axis, and two anchor nodes are to be deployed on a vertical base for y-axis [50, 54–56]. Let's consider the deployment of three anchor nodes R_1, R_2, and R_3; the unknown 2-D location of the target at any random time instance and the angles between the base and the line formed by the anchor nodes are α_{x1}, α_{x2}, α_{y1}, and α_{y2} as shown in Fig. 2.10. The nodes R_1 and R_2 form the horizontal baseline, whereas the nodes R_1 and R_3 form the vertical baseline. The unknown target location (x, y) can be computed using Eq. (2.12) as given below:

$$x = \frac{d_{ry}\sin(\alpha_{y1})\sin(\alpha_{y2})}{\sin(\alpha_{y1}+\alpha_{y2})}, \quad y = \frac{d_{rx}\sin(\alpha_{x1})\sin(\alpha_{x2})}{\sin(\alpha_{x1}+\alpha_{x2})} \tag{2.12}$$

The major drawback of angulation-based approach is the requirement of an array of at least two directional antennas to determine the angle of arrival of the RF signals. Attaching directional antennas with WSN nodes is not at all economical, and the system becomes complex.

2.5.3 *Fingerprinting*

The fingerprinting-based L&T approach involves the construction of a huge database of the RSSI measurements (also called as radio map or RF fingerprints) between the access points and the given wireless device in the off-line phase [8,

57–59]. In the online estimation phase, the new RSSI measurement vector is then compared with the stored radio map to estimate the corresponding unknown location of the target. In RSSI-based fingerprinting technique, Wi-Fi is a widely used technology because of its presence at many indoor locations and its long-range communication capability as against other technological counterparts. However, constructing a radio map is a very complex and time-consuming task. Additionally, the radio map does not give proper results, if the configuration of the indoor environment changes.

2.6 Mobility Models for Target Tracking

The target mobility model describes the movement of the target in the WSN-defined area. As the target moves, target state parameters changes, which in turn change the target state vector. In the Bayesian filtering-based target tracking, both the system model and measurement model utilize this target state vector for L&T. Various target mobility models, such as CV, CA, random walk, random waypoint, singer acceleration model, and mean-adaptive acceleration model, are discussed in tracking literature [60–65]. Out of these models, the CV and CA are more feasible to implement in WSN-based tracking.

2.6.1 *Constant Velocity (CV) Model*

In the CV model, the target state vector consists of the velocity and position of the moving target given as follows [60–63]:

$$X_k^{CV} = \left[x_k, \dot{x}_k, y_k, \dot{y}_k\right]^T \tag{2.13}$$

where $\dot{x}_k$ and x_k represent the velocity and position of target along the x-axis and y_k and $\dot{y}_k$ represent the position and velocity of target y-axis. The general state model is given as follows:

$$X_k^{CV} = \begin{bmatrix} x_k \\ \dot{x}_k \\ y_k \\ \dot{y}_k \end{bmatrix} = \begin{bmatrix} 1 & T & 0 & 0 \\ 0 & 1 & 0 & 0 \\ 0 & 0 & 1 & T \\ 0 & 0 & 0 & 1 \end{bmatrix} \begin{bmatrix} x_{k-1} \\ \dot{x}_{k-1} \\ y_{k-1} \\ \dot{y}_{k-1} \end{bmatrix} + \begin{bmatrix} \frac{1}{2}T^2 & 0 \\ 0 & 0 \\ T & \frac{1}{2}T^2 \\ 0 & 0 \\ 0 & T \end{bmatrix} \begin{bmatrix} a_k^x \\ a_k^y \end{bmatrix} \tag{2.14}$$

where a_k^x and a_k^y represent the change in velocity along x- and y-axes. The reason of saying "nearly" while describing this model is the fluctuation on velocity, which is acceleration is assumed to be Gaussian white noise.

2.6.2 Constant Acceleration (CA) Model

In the CA model, the target state vector is composed of variables such as velocity, target position, and the acceleration form given as [60–63] follows:

$$X_k^{CA} = \left[x_k, \dot{x}_k, \ddot{x}_k y_k, \dot{y}_k, \ddot{y}_k\right]^T \tag{2.15}$$

where $\ddot{x}_k$ and $\ddot{y}_k$ are the accelerations on x- and y-axes, respectively. The reason for saying "nearly" while describing this model is the fluctuation on acceleration. Let n_k^x and n_k^y be white noises. The corresponding state space model can be given as follows:

$$X_k^{CA} = \begin{bmatrix} x_k \\ \dot{x}_k \\ \ddot{x}_k \\ y_k \\ \dot{y}_k \\ \ddot{y}_k \end{bmatrix} = \begin{bmatrix} 1 & T & T^2/2 & 0 & 0 & 0 \\ 0 & 1 & T & 0 & 0 & 0 \\ 0 & 0 & 1 & 0 & 0 & 0 \\ 0 & 0 & 0 & 1 & T & T^2/2 \\ 0 & 0 & 0 & 0 & 1 & T \\ 0 & 0 & 0 & 0 & 0 & 1 \end{bmatrix} \begin{bmatrix} x_{k-1} \\ \dot{x}_{k-1} \\ \ddot{x}_{k-1} \\ y_{k-1} \\ \dot{y}_{k-1} \\ \ddot{y}_{k-1} \end{bmatrix} + \begin{bmatrix} T^2/2 & T \\ T & 1 \\ 1 & 0 \\ 0 & T^2/2 \\ 0 & T \\ 0 & 1 \end{bmatrix} \begin{bmatrix} n_k^x \\ n_k^y \end{bmatrix} \tag{2.16}$$

2.7 State Estimation Techniques for Target Tracking

The target tracking problem represents the continuous or on-demand estimation of the mobile target state during its motion in WSN by utilizing sensor measurements and system dynamic models [66–68]. Typically, the target state is composed of kinematic variables, such as the position (generally 2-D location), the velocity, and the acceleration of target [15, 19, 24]. The target state changes with respect to time as it moves around in WSN.

In the Bayesian filter-based implementations, the generalized form of target observation model and motion model can be given as follows:

$$X_k = f\left(X_{k-1}, u_{k-1}, w_{k-1}\right), \tag{2.17}$$

$$z_k = h\left(X_k\right) + v_k, \tag{2.18}$$

where X_k is the target state vector, z_k is the observation vector at the current time instance k and u_{k-1} is the control input vector. The w_{k-1} and v_k are white noise and are mutually independent.

There are various well-known Bayesian filter-based tracking algorithms, such as KF and PF in literature [15, 19, 24]. Out of them, the KF and its extensions, such as the extended Kalman filter (EKF) and the unscented Kalman filter (UKF), are widely employed. The Bayesian filter-based approach predicts the target state vector for the next time instance using the target state vector at the previous time instance using the system model, which is then corrected using the RSSI measurements collected for next time instance.

Regarding the second approach of the RSSI and Bayesian filter-based framework, the choice of PF or KF for L&T primarily depends on the amount and nature of measurement and process noise and application requirements [17, 66–68]. A detailed survey on the variety of Bayesian filter-based L&T systems is made in [66]. The survey reveals that unlike KF, PF can be applied in case of multimodal and non-Gaussian distribution, and it is superior in handling nonlinearity in measurements at the cost of high computational complexity. As WSN is memory and power constrained network, the implementation of PF on WSN for real-time target L&T applications is not suitable. As the RSSI-based target L&T system involves highly nonlinear system dynamics, the UKF shows better localization accuracy than KF and EKF.

2.7.1 Standard Kalman Filter (KF)

The KF can be applied as an estimator of the state of a dynamic system. If the system is linear and the noise in measurements is Gaussian with zero mean, then the KF is found to be the optimal estimator. The motion model and the observation model for the KF are given in Eq. (2.19) and Eq. (2.20), respectively [16, 19, 24, 25]:

$$X_k = AX_{k-1} + Bu_{k-1} + w_{k-1} \;, \tag{2.19}$$

where

- X_k is the target state vector which contains variables, such as the position, velocity, and acceleration at any time instance k.
- u_{k-1} is the vector which contains the control input.
- A is state transition matrix.
- B is the control input matrix.
- w_{k-1} is the parameter that contains the process noise values for each of the parameter in X_k.

$$z_k = H(X_k) + v_k \;, \tag{2.20}$$

where

- z_k is the vector, which contains variables, such as position, velocity, and acceleration at any time instance k.
- H is the transformation matrix.
- v_k is the vector that contains the measurement noise values for the parameter in X_k.

These two noise terms w_{k-1} and v_k are generally not dependent of each other (i.e., uncorrelated). For the CV model, the A and B matrices in Eqs. (2.19) and (2.20) can be given as follows:

$$A = \begin{bmatrix} 1 & 0 & dt & 0 \\ 0 & 1 & 0 & dt \\ 0 & 0 & 1 & 0 \\ 0 & 0 & 0 & 1 \end{bmatrix}, \; B = \begin{bmatrix} \frac{1}{2}dt^2 & 0 \\ 0 & \frac{1}{2}dt^2 \\ dt & 0 \\ 0 & dt \end{bmatrix}, \; H = I_{4\times 4} \tag{2.21}$$

The KF works in two simple phases, namely, predict and update. In the predict step, the estimate from in $k - 1$ is utilized to estimate in k. The measurements at time k are used to refine the prediction in predict step. The mathematical equations associated in predict and update steps are as follows.

Prediction

$$\bar{X}_k = A\hat{X}_{k-1} + Bu_{k-1} + w_{k-1} \; . \tag{2.22}$$

$$P_k^- = AP_{k-1}A_k^T + Q_k \; . \tag{2.23}$$

Update

$$K_k = P_k^- H_k^T \left(H_k P_k^- H_k^T + R_k \right)^{-} \; . \tag{2.24}$$

$$\hat{X}_k = \bar{X}_k + K_k \left(z_k - H_k \bar{X}_k \right) . \tag{2.25}$$

$$P_k = \left(I - K_k H_k \right) P_k^- \; , \tag{2.26}$$

where matrix K is the Kalman gain matrix and I is the identity matrix ($I_{4\times 4}$). The superscript " ^ " represents the estimate of state vector. Given the initial state vector X_{k-1} and the corresponding process covariance matrix P_{k-1}, the state vector and corresponding process covariance matrix for the next time instance k can be predicted (see Eqs. (2.22) and (2.23)). These estimates will be further updated using measurements at time instance k (see Eqs. (2.25) and (2.26)).

2.7.2 UKF

In practice, the measurement and motion models of many systems are nonlinear. For such systems, the application of standard KF is not at all useful for estimation purpose. In order to deal with this nonlinearity problem, the EKF and UKF can be employed [16, 19, 24, 25]. The EKF algorithm is very sensitive to high nonlinearities. As against the EKF, the UKF algorithm can cope up with nonlinearities by performing approximations on the target mean and covariance. The UKF algorithm is basically based on the unscented transform (UT), wherein a minimal (optimal) set of sample points (also known as sigma points) are picked around the mean deterministically.

As like the KF and UKF, the UKF operation is also based on two phases (steps), namely, predict and update. After carefully defining noise covariance matrix Q and measurement noise covariance matrix R, and initializing x and the covariance matrix P, the sigma points can be calculated as [16, 19, 24]:

$$\chi_{k-1} = \left[\hat{X}_{k-1} \quad \hat{X}_{k-1} + \gamma\sqrt{P_{k-1}} \quad \hat{X}_{k-1} + \lambda\sqrt{P_{k-1}} \right]. \tag{2.27}$$

The estimate of time step (k-1) is utilized to generate the estimate of the next time instance k in predict step (see Eqs. (2.28)–(2.33)).

Prediction

$$\chi^{*}_{k/k-1} = f\left(X_{k-1}, u_{k-1}\right) \tag{2.28}$$

$$\hat{X}_k = \sum_{i=0}^{2L} w_i^m \chi^{*}_{k/k-1} \,. \tag{2.29}$$

$$P_k = \sum_{i=0}^{2L} w_i^c [z_{i,k/k-1} - \hat{z}_k] \, [z_{i,k/k-1} - \hat{z}_k]^T + \mathrm{R}\,. \tag{2.30}$$

$$\chi_{k-1} = \left[\hat{X}_{k-1} \; \hat{X}_{k-1} + \gamma\sqrt{P_{k-1}} \; \hat{X}_{k-1} + \lambda\sqrt{P_{k-1}} \right]. \tag{2.31}$$

$$z_{k/k-1} = H\chi^{*}_{k/k-1} \,. \tag{2.32}$$

$$\hat{z}_k = \sum_{i=0}^{2L} w_i^m z_{i,k/k-1} \,. \tag{2.33}$$

In the update phase, the measurements (observations) of current time step are utilized to update this prediction to get a more accurate estimation (see Eqs. (2.34)–(2.39)).

Update

$$P_{xk,zk} = \sum_{i=0}^{2L} w_i^c \left[z_{i,k/k-1} - \hat{z}_k \right] \left[z_{i,k/k-1} - \hat{z}_k \right]^T + \mathrm{R} . \tag{2.34}$$

$$P_{xk,zk} = \sum_{i=0}^{2L} w_i^c \left[X_{i,k/k-1} - \hat{X}_k \right] \left[z_{i,k/k-1} - \hat{z}_k \right]^T + \mathrm{R} . \tag{2.35}$$

Kalman Gain

$$K_k = P_{xk,zk} \; P_{zk,zk}^{-1} . \tag{2.36}$$

Emendation State Estimate

$$\hat{x}_k = \hat{x}_{k-1} + K_k \left(z_k - \hat{z}_k \right) . \tag{2.37}$$

Error Covariance Matrix Updates

$$P_k = P_{k-1} - K_k P_{zk,zk} K_k^T , \tag{2.38}$$

where.
w_0^m is the weight of mean.
w_0^c is the weight of covariance.
λ is the scaling parameter.
L is the dimension of augmented state.

$$w_0^m = \lambda / \left(L + \lambda \right), \; w_0^c = \lambda / \left(L + \lambda \right) + \left(1 + \alpha^2 + \beta \right) , \tag{2.39}$$

where.
α is the measure of the spread of sigma points around $\hat{x}$.
α is generally the set to a small positive number, while β incorporates the prior knowledge of distribution (spread) of x.

2.8 Challenges Associated with RSSI-Based Indoor L&T

The RF-based systems can be subdivided further into Wi-Fi, Bluetooth, RFID, Zigbee, and UWB-based systems as shown in Fig. 2.11. Irrespective of the various advantages coupled with the RSSI (i.e., RF), the target tracking accuracy of the majority of existing target tracking systems is typically above 1 m, because of the various reasons as discussed below [14, 36, 39, 69–71]:

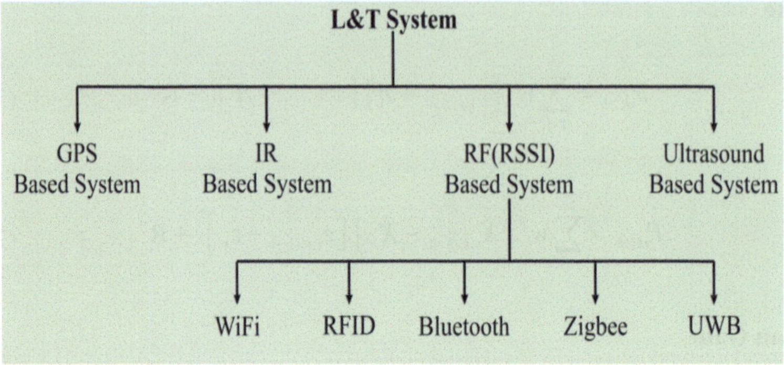

Fig. 2.11 Technology wise classification of WSN-based L&T systems

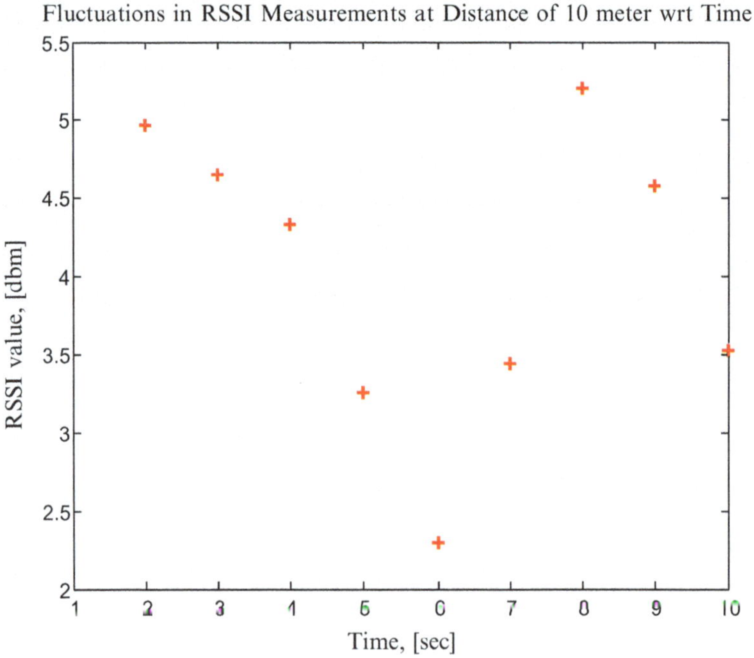

Fig. 2.12 Dynamicity in RSSI measurement

1. The RF wave moves along the different paths of varying length during its propagation and reaches at the receiver at different times with varying attenuation level because of multiple reflections. Typically, the indoor environment is generally surrounded by obstacles, and thereby, the issues related with the signal propagation, such as reflection, NLOS, and multipath propagation, degrade the target tracking accuracy in the RSSI-based system.
2. Due to the above signal propagation issues, the RSSI measurements are generally corrupted by the environmental noise. The conversion of RSSI measurements to distances requires the use of an appropriate signal path loss model as

well as its fine-tuning (wireless channel calibration) with the given environmental setup. If the wireless medium changes, then again calibration of the path loss model is necessary to accommodate these changes. In practice, the calibration of parameters of signal path loss model to account for these signal propagation issues is highly challenging.

3. Many times, the heights of transmitter and receiver nodes from the ground are different in the given wireless environment, which lead to NLOS situation. Additionally, the type of antenna and its orientation also affect the performance of RSSI-based tracking.
4. Many times, though the distance between the transmitter and the receiver as well as the transmission power and the number of obstacles between the transmitter and the receiver is kept constant, the received signal strengths are found to be varying drastically. For instance, for a distance of 10 m between the transmitter and the receiver, the RSSI noted down is fluctuating drastically as shown in Fig. 2.12.
5. In case of range-based target tracking system, the RSSI measurements are converted into distances. It is challenging to deal with a highly nonlinear RSSI-distance relationship.
6. The nature of RSSI measurements is environment dependent. This dependence is especially more pronounced in an indoor environment. For instance, a little change in the position of transmitter or receiver in the given indoor environment, the RSSI values change drastically. Thus, even for the same environmental setup, there are lesser chances of repeatability and regularity in the RSSI values. In other words, the RSSI signal is not periodic.

References

1. G. Xu, W. Shen, X. Wang, Applications of wireless sensor networks in marine environment monitoring: a survey. Sensors (Switzerland) **14**(9), 16932–16954 (2014). https://doi.org/10.3390/s140916932
2. P. Kumar, H.J. Lee, Security issues in healthcare applications using wireless medical sensor networks: a survey. Sensors **12**(1), 55–91 (2012). https://doi.org/10.3390/s120100055
3. B. Rashid, M.H. Rehmani, Applications of wireless sensor networks for urban areas: a survey. J. Netw. Comput. Appl. **60**, 192–219 (2016). https://doi.org/10.1016/j.jnca.2015.09.008
4. I.F. Akyildiz, T. Melodia, K.R. Chowdhury, A survey on wireless multimedia sensor networks. Comput. Netw. **51**(4), 921–960 (2007). https://doi.org/10.1016/j.comnet.2006.10.002
5. I. Khemapech, I. Duncan, A. Miller, A survey of wireless sensor networks technology, in *6th Annual Postgraduate Symposium on the Convergence* of *Telecommunications, Networking and Broadcasting*, vol. 6 (2005)
6. W. Dargie, C. Poellabauer, *Fundamentals of Wireless Sensor Networks: Theory and Practice* (Wiley, New York, 2011). https://doi.org/10.1002/9780470666388
7. N. Patwari, J.N. Ash, S. Kyperountas, A.O. Hero, R.L. Moses, N.S. Correal, Locating the nodes: Cooperative localization in wireless sensor networks. IEEE Signal Process. Mag. **22**(4), 54–69 (2005). https://doi.org/10.1109/MSP.2005.1458287
8. R. Faragher, R. Harle, Location fingerprinting with bluetooth low energy beacons. IEEE J. Sel. Areas Commun. (2015). https://doi.org/10.1109/JSAC.2015.2430281
9. Y. Miao, H. Wu, L. Zhang, The accurate location estimation of sensor node using received signal strength measurements in large-scale farmland. J. Sensors (2018). https://doi.org/10.1155/2018/2325863

10. M. Perkins, N. Patwari, A.O. Hero, R.J. O'Dea, N.S. Correal, Relative location estimation in wireless sensor networks. IEEE Trans. Signal Process. (2003). https://doi.org/10.1109/tsp.2003.814469
11. S.R. Jondhale, R.S. Deshpande, S.M. Walke, A.S. Jondhale, Issues and challenges in RSSI based target localization and tracking in wireless sensor networks, in *2016 International Conference on Automatic Control and Dynamic Optimization Techniques (ICACDOT)*, (2017). https://doi.org/10.1109/ICACDOT.2016.7877655
12. A. PAL, Localization algorithms in wireless sensor networks: current approaches and future challenges. Netw. Protoc. Algorithms (2011). https://doi.org/10.5296/npa.v2i1.279
13. X. Zheng, H. Liu, J. Yang, Y. Chen, R.P. Martin, X. Li, A study of localization accuracy using multiple frequencies and powers. IEEE Trans. Parallel Distrib. Syst. (2014). https://doi.org/10.1109/TPDS.2013.281
14. F. Viani, P. Rocca, G. Oliveri, D. Trinchero, A. Massa, Localization, tracking, and imaging of targets in wireless sensor networks: An invited review. Radio Sci. (2011). https://doi.org/10.1029/2010RS004561
15. S. R. Jondhale, R. S. Deshpande, Tracking target with constant acceleration motion using Kalman filtering, in *2018 International Conference On Advances in Communication and Computing Technology (ICACCT)* (2018). https://doi.org/10.1109/ICACCT.2018.8529628
16. S. R. Jondhale, R. S. Deshpande, Modified Kalman filtering framework based real time target tracking against environmental dynamicity in wireless sensor networks. Ad Hoc Sens. Wirel. Netw. **40**, 119–143 (2018)
17. A.S. Paul, E.A. Wan, RSSI-based indoor localization and tracking using sigma-point kalman smoothers. IEEE J. Sel. Top. Signal Process. (2009). https://doi.org/10.1109/JSTSP.2009.2032309
18. S. Mahfouz, F. Mourad-Chehade, P. Honeine, J. Farah, H. Snoussi, Target tracking using machine learning and kalman filter in wireless sensor networks. IEEE Sensors J. (2014). https://doi.org/10.1109/JSEN.2014.2332098
19. S.R. Jondhale, R.S. Deshpande, Kalman filtering framework based real time target tracking in wireless sensor networks using generalized regression neural networks. IEEE Sensors J. **19**(1), 224–233 (2019)
20. L. Gui, T. Val, A. Wei, R. Dalce, Improvement of range-free localization technology by a novel DV-hop protocol in wireless sensor networks. Ad Hoc Netw **24**(Part B), 55–73 (2015). https://doi.org/10.1016/j.adhoc.2014.07.025
21. J.L.M. Garcia, J. Tomas, F. Boronat, The development of two systems for indoor wireless sensors self-location. Ad Hoc Sensor Wirel. Netw. **8**(3–4), 235–258 (2009)
22. B. Wagner, D. Timmermann, G. Ruscher, T. Kirste, Device-free user localization utilizing artificial neural networks and passive RFID, in *2012 Ubiquitous Positioning, Indoor Navigation, and Location Based Service (UPINLBS)* (2012). https://doi.org/10.1109/UPINLBS.2012.6409762
23. F. Darakeh, G. R. Mohammad-Khani, P. Azmi, An accurate distributed rage free localization algorithm for WSN, in *2017 Iranian Conference on Electrical Engineering (ICEE)* (2017). https://doi.org/10.1109/IranianCEE.2017.7985388
24. S. R. Jondhale, R. S. Deshpande, GRNN and KF framework based real time target tracking using PSOC BLE and smartphone. Ad Hoc Netw. (2019). https://doi.org/10.1016/j.adhoc.2018.09.017
25. S. Jondhale, R. Deshpande, Self recurrent neural network based target tracking in wireless sensor network using state observer. Int. J. Sensors Wirel. Commun. Control (2018). https://doi.org/10.2174/2210327908666181029103202
26. K. Heurtefeux, F. Valois, Is RSSI a good choice for localization in wireless sensor network?, in *2012 IEEE 26th International Conference on Advanced Information Networking and Applications* (2012). https://doi.org/10.1109/AINA.2012.19
27. S. Kumar, S. Lee, Localization with RSSI values for wireless sensor networks: an artificial neural network approach. Int. J. Comput Netw. Commun. **8**, 61–71 (2014). https://doi.org/10.3390/ecsa-1-d007
28. L. Mihaylova, D. Angelova, D.R. Bull, N. Canagarajah, Localization of mobile nodes in wireless networks with correlated in time measurement noise. IEEE Trans. Mob. Comput. (2011). https://doi.org/10.1109/TMC.2010.132

29. A. Zanella, Best practice in RSS measurements and ranging. IEEE Commun. Surv. Tutorials (2016). https://doi.org/10.1109/COMST.2016.2553452
30. A. T. Parameswaran, M. I. Husain, S. Upadhyaya, Is RSSI a reliable parameter in sensor localization algorithms - an experimental study. IEEE Int. Symp. Reliab. Distrib. Syst. (2009)
31. W. Rong-Hou, L. Yang-Han, T. Hsien-Wei, J. Yih-Guang, C. Ming-Hsueh, Study of characteristics of RSSI signal, in *2008 IEEE International Conference on Industrial Technology* (2008). https://doi.org/10.1109/ICIT.2008.4608603
32. Q. Dong, W. Dargie, Evaluation of the reliability of RSSI for indoor localization, in *2012 International Conference on Wireless Communications in Underground and Confined Areas* (2012). https://doi.org/10.1109/ICWCUCA.2012.6402492
33. F. Zafari, I. Papapanagiotou, M. Devetsikiotis, T. J. Hacker, Enhancing the accuracy of iBeacons for indoor proximity-based services, in *2017 IEEE International Conference on Communications (ICC)* (2017). https://doi.org/10.1109/ICC.2017.7996508
34. T. K. Sarkar, Z. Ji, K. Kim, A. Medouri, M. Salazar-Palma, A Survey of various propagation models for mobile communication. IEEE Antennas Propag. Mag. (2003). https://doi.org/10.1109/MAP.2003.1232163
35. S. Vougioukas, H.T. Anastassiu, C. Regen, M. Zude, Influence of foliage on radio path losses (PLs) for wireless sensor network (WSN) planning in orchards. Biosyst. Eng. (2013). https://doi.org/10.1016/j.biosystemseng.2012.08.011
36. S.S. Saad, Z.S. Nakad, A standalone RFID indoor positioning system using passive tags. IEEE Trans. Ind. Electron. (2011). https://doi.org/10.1109/TIE.2010.2055774
37. J. Zhu, H. Xu, Review of RFID-based indoor positioning technology, in *Innovative Mobile and Internet Services in Ubiquitous Computing. IMIS 2018. Advances in Intelligent Systems and Computing*, (Springer, Cham, 2019). https://doi.org/10.1007/978-3-319-93554-6_62
38. M. Bouet, A. L. Dos Santos, RFID tags: positioning principles and localization techniques, in *2008 1st IFIP Wireless Days* (2008). https://doi.org/10.1109/WD.2008.4812905
39. M.M. Soltani, A. Motamedi, A. Hammad, Enhancing cluster-based RFID tag localization using artificial neural networks and virtual reference tags. Autom. Constr. (2015). https://doi.org/10.1016/j.autcon.2015.03.009
40. D. Fox, M. Philipose, D. Hahnel, W. Burgard, K. Fishkin, Mapping and localization with RFID technology, in *IEEE International Conference on Robotics and Automation, 2004. Proceedings. ICRA '04* (2004). https://doi.org/10.1109/robot.2004.1307283
41. H. D. Chon, S. Jun, H. Jung, W. An, Using RFID for accurate positioning. J. Glob. Position. Syst. (2010). https://doi.org/10.5081/jgps.3.1.32
42. S. L. Ting, S. K. Kwok, A. H. C. Tsang, G. T. S. Ho, The study on using passive RFID tags for indoor positioning. Int. J. Eng. Bus. Manag. **3**, 9–15 (2011)
43. Y. Zhang, X. Li, M. Amin, Principles and techniques of RFID positioning, in *RFID Systems: Research Trends and Challenges*, (2010)
44. C.E.T. Galvan, I. Galvan-Tejada, E.I. Sandoval, R. Brena, Wifi bluetooth based combined positioning algorithm. Proc. Eng. **35**, 101–108 (2012). https://doi.org/10.1016/j.proeng.2012.04.170
45. X. Zhao, Z. Xiao, A. Markham, N. Trigoni, Y. Ren, Does BTLE measure up against WiFi? A comparison of indoor location performance, in *European Wireless 2014; 20th European Wireless Conference* (2014)
46. Y. S. Chiou, C. L. Wang, S. C. Yeh, M. Y. Su, Design of an adaptive positioning system based on WiFi radio signals. Comput. Commun. (2009). https://doi.org/10.1016/j.comcom.2009.04.003
47. Y. W. Prakash, V. Biradar, S. Vincent, M. Martin, A. Jadhav, Smart bluetooth low energy security system, in *2017 International Conference on Wireless Communications, Signal Processing and Networking (WiSPNET)* (2018). https://doi.org/10.1109/WiSPNET.2017.8300139
48. R. Faragher, R. Harle, An analysis of the accuracy of bluetooth low energy for indoor positioning applications, in *International Technical Meeting of the Satellite Division of the Institute of Navigation* (2014). https://doi.org/10.1016/j.ajpath.2011.02.025
49. Y. Zhuang, J. Yang, Y. Li, L. Qi, N. El-Sheimy, Smartphone-based indoor localization with bluetooth low energy beacons. Sensors (Switzerland) (2016). https://doi.org/10.3390/s16050596

50. A. De Blas, D. López-de-Ipiña, Improving trilateration for indoors localization using BLE beacons, in *International Multidisciplinary Conference on Computer and Energy Science* (2017)
51. E. MacKensen, M. Lai, T. M. Wendt, Bluetooth Low Energy (BLE) based wireless sensors, in *2012 IEEE Sensors* (2012). https://doi.org/10.1109/ICSENS.2012.6411303
52. M. S. Pan, Y. C. Tseng, ZigBee wireless sensor networks and their applications. Sens. Networks Config. Fundam. Stand. Platforms Appl. (2007). https://doi.org/10.1007/3-540-37366-7_16
53. R. Mardeni and S. Nizam, .Node positioning in zigbee network using trilateration method based on the received signal strength indicator (RSSI). Eur. J. Sci. Res.(2010)
54. Z. Yang, Y. Liu, X. Y. Li, Beyond trilateration: On the localizability of wireless ad hoc networks. IEEE/ACM Trans. Netw. (2010). https://doi.org/10.1109/TNET.2010.2049578
55. F. Thomas, L. Ros, Revisiting trilateration for robot localization. IEEE Trans. Robot. (2005). https://doi.org/10.1109/TRO.2004.833793
56. J. Uren, W.F. Price, Triangulation and trilateration, in *Surveying for Engineers*, (2015)
57. M. Stella, M. Russo, D. Begusic, Location determination in indoor environment based on RSS fingerprinting and artificial neural network, in *2007 9th International Conference on Telecommunications* (2007). https://doi.org/10.1109/CONTEL.2007.381886
58. Y. Zhuang, Y. Li, L. Qi, H. Lan, J. Yang, N. El-Sheimy, A two-filter integration of MEMS sensors and WiFi fingerprinting for indoor positioning. IEEE Sensors J. (2016). https://doi.org/10.1109/JSEN.2016.2567224
59. S. He, S. H. G. Chan, Wi-Fi fingerprint-based indoor positioning: Recent advances and comparisons. IEEE Commun. Surv. Tutorials. (2016). https://doi.org/10.1109/COMST.2015.2464084
60. R. A. Pushpa, A. Vallimayil, V. R. S. Dhulipala, Impact of mobility models on mobile sensor networks, in *2011 3rd International Conference on Electronics Computer Technology* (2011). https://doi.org/10.1109/ICECTECH.2011.5941866
61. T. Camp, J. Boleng, V. Davies, A survey of mobility models for ad hoc network research. Wirel. Commun. Mob. Comput. (2002). https://doi.org/10.1002/wcm.72
62. R. Silva, J. Sa Silva, F. Boavida, Mobility in wireless sensor networks - survey and proposal. Comput. Commun. (2014). https://doi.org/10.1016/j.comcom.2014.05.008
63. L. Mihaylova, D. Angelova, S. Honary, D. R. Bull, C. N. Canagarajah, B. Ristic, Mobility tracking in cellular networks using particle filtering. IEEE Trans. Wirel. Commun. (2007). https://doi.org/10.1109/TWC.2007.05912
64. A. U. R. Khan, S. Ali, S. Mustafa, M. Othman, Impact of mobility models on clustering based routing protocols in mobile WSNs, in *2012 10th International Conference on Frontiers of Information Technology* (2012). https://doi.org/10.1109/FIT.2012.72
65. S. Jardosh, P. Ranjan, A survey: topology control for wireless sensor networks, in *2008 International Conference on Signal Processing, Communications and Networking* (2008). https://doi.org/10.1109/ICSCN.2008.4447231
66. D. Fox, J. Hightower, L. Liao, D. Schulz, G. Bordello, Bayesian filtering for location estimation. IEEE Pervasive Comput. (2003). https://doi.org/10.1109/MPRV.2003.1228524
67. F. Zafari, I. Papapanagiotou, T. J. Hacker, A novel Bayesian filtering based algorithm for rssi-based indoor localization, in *2018 IEEE International Conference on Communications (ICC)* (2018). https://doi.org/10.1109/ICC.2018.8423012
68. M.S. Arulampalam, S. Maskell, N. Gordon, T. Clapp, A tutorial on particle filters for online nonlinear/nongaussian bayesian tracking, in *Bayesian Bounds for Parameter Estimation and Nonlinear Filtering/Tracking*, (2007)
69. D. Balachander, T. R. Rao, G. Mahesh, RF propagation investigations in agricultural fields and gardens for wireless sensor communications, in *2013 IEEE Conference on Information & Communication Technologies* (2013). https://doi.org/10.1109/CICT.2013.6558195
70. M. Malajner, K. Benkič, P. Planinšič, Ž. Čučej, The accuracy of propagation models for distance measurement between WSN nodes, in *2009 16th International Conference on Systems, Signals and Image Processing* (2009). https://doi.org/10.1109/IWSSIP.2009.5367782
71. H. Liu, H. Darabi, P. Banerjee, J. Liu, Survey of wireless indoor positioning techniques and systems. IEEE Trans. Syst. Man Cybern. Part C Appl. Rev. (2007). https://doi.org/10.1109/TSMCC.2007.905750

Chapter 3
Survey of Existing RSSI-Based L&T Systems

3.1 Survey of Application of Various Wireless Technologies for Indoor Tracking

The RF-based indoor target L&T can be achieved with the help of a wide range of wireless technologies, such as RFID, Wi-Fi, Bluetooth, Zigbee, UWB, and ultrasound. The pros and cons of these technologies are presented in Table 1 [1–5].

The growing commercial interest in indoor LBS has led to the development of many indoor positioning techniques. Out of all of these technological alternatives, Wi-Fi has emerged as a key technology due to the already existing wireless LANs [6–8]. Today, almost all the present smartphones, laptops, as well as many other portable user electronic devices are Wi-Fi enabled. Therefore, the Wi-Fi has been used by many researchers for indoor L&T. In Wi-fi based, the existing Wi-Fi APs are utilized as reference points to generate RSSI measurements. That means, the L&T system can be developed without the need for extra infrastructure. The major drawback of Wi-Fi-based L&T systems are as follows:

(i) The Wi-Fi infrastructure is used for network communication. Hence, it is not useful for the localization problem outside the given environment.
(ii) Additionally, it is associated with very high power consumption.
(iii) Uncontrolled interference in the ISM band can significantly affect the L&T accuracy.

Overall, the positioning accuracy that can be achieved in the Wi-Fi-based indoor localization is only around 2.0–2.5 m. Thus, even though sensor nodes are provided with Wi-Fi capability, due to various disadvantages associated with it, Wi-Fi-based L&T is not at all an economical option.

The recent introduction of the BLE radio protocol has gained a lot of attention in indoor L&T applications. The low-cost and easily deployable BLE-based nodes

S. R. Jondhale et al., *Received Signal Strength Based Target Localization and Tracking Using Wireless Sensor Networks*, EAI/Springer Innovations in Communication and Computing, https://doi.org/10.1007/978-3-030-74061-0_3

Table 1 Comparison of various wireless technologies for target L&T

Wireless technology	Frequency of operation	Communication range	Pros	Cons
Wi-Fi	900 MHz, 3.6 GHz, 2.4 GHz, 4.9 GHz, 5 GHz	50–100 m	Readily available in offices, campuses, wireless devices	High power consumption, low accuracy around 2 m due to signal fluctuations
RFID	13.56 MHz	5–6 m	No need of LOS	Short range, moderate accuracy
Bluetooth low energy (BLE)	2.4 GHz/5 GHz	30 m	Low power consumption, readily available, long range	Suitable for low data rate (1 Mbps and 2 Mbps)
Zigbee	2.4 GHz	30 m	Low power consumption	Not available in portable instruments
UWB	3.1–10.6 GHz	70 m	Low power consumption, high resolution	High cost, not available in portable instruments
Ultrasound	20 Hz–20 KHz	6–9 m	High accuracy	High cost, not scalable

(beacons) have many distinct advantages over Wi-Fi, such as availability on numerous portable devices such as smartphone and laptops. The two major BLE-based protocols primarily used for LBS are iBeacons and Eddystone as discussed in the previous chapter. Against the old Bluetooth standard, these newer BLE technologies can give a communication range of 70–100 m and data rate of 24 Mbps with low power consumption. Although the BLE nodes may be used with various different measurements, such as AoA, RSSI, and ToF, many existing BLE-based L&T implementations rely on RSSI. However, BLE-based L&T is generally coupled with a high localization accuracy in the case of RSSI measurement type. For instance, in [8] the detailed study of BLE fingerprinting-based localization system is carried out using 19 beacon nodes distributed around 600 square meters indoor environment to locate the moving consumer object. This work also investigates the impact of transmission power, beacon density, and transmission frequency on localization accuracy. The authors have proved that the RSSI measurements obtained with BLE devices are more stable than that with Wi-Fi devices. Because of on-chip antenna and transceiver, these tiny BLE-enabled devices can be utilized as WSN nodes. However, care has to be taken against indoor issues related to signal propagation, such as multipath propagation, fading, and NLOS, while designing target L&T algorithms for BLE-based implementations.

Another important low-cost wireless technology used for indoor L&T is Zigbee. Zigbee has the capability of low data rate and reduced power consumption. Its network layer takes care of multi-hop communication and network management, whereas its application layer handles the development of the application. By attaching the target with a Zigbee node, the RSSI metric can be used to estimate the location of the tagged target with the help of other deployed nodes. The major drawback of Zigbee-based network is its unavailability with most of the user devices; therefore, Zigbee is not preferred for indoor L&T. In the Zigbee-based system in [9], the RSSI is converted to the distance, which is then fed to the trilateration method for location estimation. The location estimations from trilateration are mapped using Google maps. Another low-cost alternative RFID has the capability to transfer and store data using RF waves from the transmitter to any circuit, which has compatibility of RF. The RFID tag can also be attached with a target for its automatic tracking. These systems consist of an RFID reader, which can communicate with the attached RFID tag. The basic categories of RFID are as follows: active RFID and passive RFID. Although active RFIDs can be used for L&T, it is not available on many portable user gadgets. The passive RFID is limited in the communication range; they are popular in proximity-based services. Apart from the short communication range, it has moderate accuracy.

In UWB-based systems, short-period pulses with a very low duty cycle are transmitted in the frequency range of 3.1–10.6 GHz. The short-range communication and low power consumption capability as well as high-resolution capability in the time domain make it a good option for wireless-based L&T. It can deal with multipath propagation issues without the need for complex estimation algorithms to provide an accurate localization. However, the measurement metric in UWB-based system is ToA or TDoA. Similar to UWB, the acoustic signal and ultrasound also use ToA or TDoA for L&T applications. We know that the microphone sensor is always equipped with smartphones. In the acoustic signal-based system, the microphone sensor is utilized to capture the acoustic signal emitted from various sound sources to compute location, by utilizing ToF concept. The ultrasound-based system uses sound velocity and ultrasound signals to estimate distance between the transmitter node and the receiver node. Unlike RF signals, the biggest disadvantage of the ultrasound-based system is the significant variations in the sound velocity, especially when temperature and humidity of the environment change. In order to cope up with this limitation, temperature sensors are usually deployed along with the ultrasound systems. As the aim of the proposed research work is to design RSSI-based, the discussion of UWB and acoustic signal-based localization technique is beyond the scope of this thesis.

3.2 Survey of Application of Bayesian Filtering in RSSI-Based Target Tracking

From the RSSI-based tracking system in [10], a novel optimized transmit power level strategy is presented. In this approach, the WSN is deployed in close proximity to the mobile target. The optimal transmission power level strategy enables a perfect conversion of the RSSI measurements into distances. This WSN-based implementation also adopts advanced signal processing techniques to address problems, such as channel distortion and packet loss. The processing stage preprocesses the RSSI measurements to reduce the fluctuations in it, and then advanced filtering is applied to obtain the location estimates. Though the proposed system demonstrates a very high tracking accuracy of 4.33 cm, the considered indoor area for tracking experiments is only 80 cm × 80 cm, and the use of high transmission power is not suitable for WSN-based implementations.

As the target tracking problem involves recursive estimation of the target locations during its motion, the Bayesian framework is a suitable algorithmic option. It is basically a probability-based framework, in which estimates of the unknown state are obtained sequentially with the help of noisy observations by adopting the observation likelihood and dynamic predictive model. PF, KF, EKF, and UKF are the popular Bayesian filters in RSSI-based L&T. In [11, 12], the authors have carried out a survey of Bayesian filter-based implementations for L&T with the help of infrared, ultrasound, as well as laser range finders. This survey concludes that though for non-Gaussian and multimodal cases the PF algorithm can converge to the true posterior state distribution, the KF algorithms are more efficient in the context of memory usage and computational capability.

PF has been used successfully for real-time target tracking scenarios. The work in [13] discusses the PF-based L&T of moving target using RSSI values with measurement noises correlated in time. The authors propose novel multimodal auxiliary PF implementations to cope up with measurement noises correlated in time. The presented algorithm is verified using simulated and real RSSI measurements, and it yields high localization accuracy. However, due to PF, the computational complexity of the proposed tracking system is higher. The PF-based system described by [14] discusses human motion model with random acceleration, whereas the observation model is based on noisy RSSI measurements. The PF-based work in [15] discusses the application of fingerprinting technique on the Wi-Fi signal-based propagation model for L&T. However, the proposed method is based on the assumption of the attenuation of RF signal with radial symmetry as well as the requirement of huge training dataset for creating RF fingerprint of the given wireless environment.

KF methods have long been popular for real-time L&T and navigation applications. In [16], the authors have presented a combination of EKF and cooperative tracking system to obtain location estimates using RSSI measurements. The result showed a reduction in error by 47.47% as compared to cooperative tracking system alone and mean square error of 1.09 m. However, the computational time of the proposed system is 120.74 s, while that of cooperative tracking alone is 41.73 s. That means, although the system tracking accuracy is high, the real-time performance is very slow to provide location estimates. The work in [17] is based on hybrid KF-based system to locate and track the moving object using popular cricket localization system in the indoor environment. The experimental results yield good performance of the proposed approach over the other methods. Few implementations in the literature also exploit the advantages of both KF and PF. In [18], PF-EKF-based cascaded algorithm is discussed. The objective of the combination of EKF and PF is to lower down the effect of multipath propagation and noisy RSSI. The simulation results prove that the proposed approach improvises the tracking accuracy by 33.9% and 31.3% in the 2-D and 3-D environments, respectively, as against PF-based implementation alone for the same environmental setting. The localization accuracy obtained is 0.97 m. In [19], the authors use a cascading of KF with PF for indoor target tracking application using iBeacon nodes. The proposed system improved localization accuracy by 28.16% and 25.59% in 2-D and 3-D localization, respectively, as compared to the PF-based algorithm alone.

The smartphone can also be used to assist the KF framework in indoor target L&T. In [20], the authors have proposed a smartphone-based sensor fusion-based KF framework by combining Wi-Fi and PDR. As the overall proposed system runs on the smartphone, the problem of the sensor fusion is formulated in a linear regression manner. The proposed implementation provides localization accuracy of around 1 m. From the research work in [21], a smartphone-based indoor L&T system using iBeacon nodes has been proposed. In the experimentation, key issues, namely, estimation of walking direction and step detection, are investigated. The PDR-based approach suffers from the drift in walking distance. In order to address this problem, the iBeacon is used to calibrate the PDF drift. By studying the measurements from iBeacon nodes, the authors propose an efficient calibration range, where EKF can be applied. The experimental results demonstrated the effectiveness of the proposed approach in terms of improved localization accuracy.

The research in [22] investigates the application of KF, dead reckoning, and trilateration for indoor target tracking. In this, the BLE node broadcast RF signal and the RSSI values are measured using a smartphone. To filter the noise from RSSI measurements, KF is used, whereas target localization is performed with the help of trilateration and dead reckoning techniques. The research findings are that fusion based on KF and trilateration has good tracking performance as compared to that from separate implementations of trilateration, dead reckoning, and KF. Though the proposed system yields localization accuracy below 0.75 m, the indoor layout area considered is very small. Also, as large number of BLE modules is deployed at very smaller distances from each other, the overall system cost is very high. The research work in [23] employs EKF and circularly polarized (CP) antennas for indoor tracking problem using RSSI measurements. The EKF is fed with robot motion model with the CV model, and both the anchor node and the mobile node are equipped with CP antennas for information exchange. The CP antenna maintains LOS communication between nodes, such that the RSSI measurements are accurate and stable. Four anchor nodes are kept at corners of 4 m × 4 m area, and the robot movement is assumed to be along a predefined trajectory within the defined area. The proposed system demonstrates a very high tracking accuracy with the maximum localization error of 0.52 m. The major drawback of the proposed system is the need of expensive CP antennas, and the defined area is too small to prove the effectiveness of the approach for the large-scale indoor environment. The indoor localization using Wi-Fi RSSI measurements is discussed in [24]. To enhance the target localization accuracy, virtual access points (VAP) are added in the system using linear regression statistical model, by exploiting the correlation between existing APs. That means, the density of APs is increased without actually adding a new hardware in the system. The objective of the application of KF is to reduce the impact of noise in the RSSI measurements. However, the localization error observed of the proposed system is very high (i.e., 4.49 m).

The authors in [25] have proposed the sigma-point Kalman smoother (SPKS)-based L&T system, which is based on the fusion of a dynamic human walking model and series of sensor measurements. The tracking system utilizes the measurements from Wi-Fi APs (RSSI measurements), binary foot switches, and binary IR motion sensors. The target carries a small wearable device, which measures the RSSI from at least three Wi-Fi APs that are kept at predefined locations. The dynamic model considered in this work is fused with a room wall model. Instead of using a fixed path loss model, the proposed system learns the RSSI-location relation during the training system. The proposed system demonstrates high localization accuracy in the scale of few meters. The work in [26] presents current-statistical model-based adaptive UKF (CAUKF) for maneuvering target tracking problem using RSSI measurements. The relation between the target motion and the corresponding positions of the neighboring nodes is utilized to carry out the adaptation in the UKF process noise covariance matrix. Speaking in more clear words, a novel modified Sage-Husa estimator is applied to adapt the process noise covariance matrix, whereas the UKF adaptive covariance matrix for measurement noise is implemented by a predefined fuzzy rule set. The simulation results show low system latency and better tracking accuracy. Another data fusion-based work is presented in [27]. This work fuses the RSSI measurements from WLAN and target velocity information from IMU module using UKF to estimate 2-D locations of moving node in an indoor scenario. The mobile unit is supposed to be equipped with IMU and a wireless transceiver. Although the collected RSSI values are corrupted with multipath fading and quantization noise, the proposed strategy achieves RMSE around 1 m.

3.3 Survey of Application of ANN in RSSI-Based Target Tracking

Though the Bayesian filter-based approach can provide improved performance than the traditional methods, the system models in the tracking systems are generally nonlinear and have some degree of mismatch with actual reality. Therefore, it is very difficult to achieve higher tracking accuracy with the help of the Bayesian framework alone. Therefore, in the indoor applications involving dynamicity in RSSI measurements, some advanced signal processing techniques, which can filter out the noise in the measurements, are needed. As the artificial neural network (ANN) does not require the prior knowledge of noise distribution, the application of a suitable ANN has a huge potential in the RSSI based L&T using WSN. Unlike the KF, the ANN can be utilized to model almost any complicated relationships easily.

Very few existing localization systems are based on the application of the ANN. Especially in the localization systems wherein Bluetooth technology is used, the ANN has not been employed. Basically, the ANN is a universal approximator, which has the capability of mapping functions in the multivariable spaces. The capability to quickly learn and generalize is considered as one of the key advantages of the ANN. The success of ANN-based target L&T system depends on the parameters, namely, number of hidden layers in ANN, node density per layer, and the selection of the transfer function. In the previous ANN-based L&T systems, these parameters are chosen based on merely the designer's experience, and thus it cannot be directly applied to future L&T systems as it is.

There are a wide variety of ANN structures, which can be applied for the problem of target L&T. For instance, the multilayer perceptron (MLP) can be used effectively for L&T problem. The MLP is basically a multilayer feed-forward ANN, which has the capability of mapping any given set of inputs to the corresponding target output sets. These architectures are based on supervised learning and can be used to address the problem of the noise uncertainty involved in RSSI measurements for the given indoor environment. In [28], the authors investigated the tracking of moving wireless sensors in a given indoor setup, which is uncertain and harsh. In this work, two MLP-based localization algorithms are discussed and are compared with the traditional trilateration-based estimation. The results obtained with the proposed MLP-based algorithms with two hidden layers show superior performance than trilateration. At the end, the real-time experiments are conducted to verify the proposed algorithm by deploying the cricket motes. The work in [29] investigated the indoor target localization with three types of ANN architectures, namely, radial basis function (RBF), MLP, and recurrent neural networks (RNN). The data for training is collected by keeping the anchor node at each of the corners of 300 cm × 300 cm area and collecting RSSI measurements at the sensor node moving through this area. The proposed ANN is trained with these multiple sets of fluctuating RSSI measurements of 70 target locations. Once the ANNs are trained with the collected database, the testing RSSI datasets are applied by moving the node through the deployed WSN by following a predefined trajectory. Additionally, apart from the comparison of localization performances of these three ANN architectures, the localization performance of the standard KF-based implementation is also recorded. The carried-out experimentation concludes that despite the fact that RBF has large computational complexity, it shows the highest localization accuracy than the rest of the other strategies. The KF displays relatively large errors than all the ANN-based implementations, especially at the boundaries of the WSN area. The work in [30] presents an ANN-based approach to locate the target in WSN using ANN structures and genetic algorithm with the help of RSSI measurements. The simulation experiments are carried out using MATLAB to collect data for ANN input for indoor environment of 26 m × 26 m with an anchor density of 8. The proposed algorithm yields an RMSE error of 0.41 m with minimum and maximum localization errors of 1.07 m and 0.014 m, respectively. In [31], the MLP network is used for RF fingerprinting-based target localization using deployed WSN. The efficacy of this proposed work is evaluated using 13 back propagation training algorithms. An ensemble consisting of four MLP networks with a different capacity of inputs is proposed in [32]. In the proposed approach, ANN architecture with four inputs is used for the L&T. The results show that the proposed algorithm is superior to GA- and fuzzy-based implementation in the context of localization accuracy. The distinct drawback of this implementation is that it is not at all a scalable implementation. Two ANN algorithms, namely, Bayesian regularization and gradient descent are used by [33] to compute mobile sensor node location in the given indoor environment. Two measurement metrics, namely, link quality indicator (LQI) and RSSI, received using Zigbee module are utilized to train the proposed ANN-based localization algorithm. The experimentation is tested with each single measurement metric alone as well as with the combination of both. From the obtained results, the author concludes that RSSI or LQI alone is not enough to locate the mobile sensor nodes accurately. In more specific words, if the combination of these two metrics is used, then the localization accuracy can be improved successfully. The localization accuracy obtained with the proposed strategy is 1.65 m.

The authors in [34] developed a WSN-based indoor localization system by adopting on RSSI fingerprinting and ANN. The ANN formulates the relationship between different RSSI and mobile node locations in the given indoor environment. This ANN-based system yields the average localization error of 1.79 m. In [32], the authors proposed two RSSI-based range-free localization systems. The first approach employs a fuzzy logic system to locate nodes using the summation of edge weights of each deployed anchor node. These optimal edge weights are determined using the designed genetic algorithm. The second implementation is based on ANN which takes RSSI measurements as the input. The implementation in [35] proposes an ANN framework trained with LVQ to solve the problem of indoor sensor node localization. The ANN framework takes RSSI as the input, whereas it gives the location area of the sensor node in the given indoor environment as the output. However, this approach is particularly not useful in applications wherein the exact target position is required.

Yet another novel combination of machine learning and KF is presented in [36] to estimate instantaneous positions of a mobile target with the help of 20 static sensor nodes. This work constructs the radio map of the indoor environment of 100 m × 100 m area with the help of RSSI measurements. This RF fingerprint along with machine learning algorithm is utilized to compute the first location estimate of the mobile target using only the RSSI information. The prediction target locations are obtained by utilizing the target acceleration information. The first location estimates and the predicted locations are fused together with the help of KF to refine the location estimates further. The proposed system demonstrates the tracking accuracy around 1 m even under noisy acceleration information and RSSI measurements. In [37], the feed-forward MLP-based approach for localization, using the RSSI values from the anchor nodes, is presented. The anchor nodes are deployed at each corner of 5 m × 4 m indoor environment. This work also investigates the efficacy of five different training algorithms to train MLP. For the validation of the proposed algorithm, the MLP is implemented with the Arduino programming language on the hardware, such that it consists of a 12-12-2 structure with 4 input nodes, 24 nodes, and 2 nodes in two hidden and the output layers, respectively, and is trained using the Bayesian regulation learning algorithm. After the training phase, the proposed implementation receives the RSSI values using anchor nodes as an input at a particular time instance, and it provides the corresponding 2-D location of target node at that instance of time. The proposed scheme provides the average localization error of 30 cm for the unknown dataset of RSSI measurements for the given environmental setup. In [38], three ANN architectures, namely MLP, RBF, and fully connected neural networks (FCNN), are implemented for target localization in the area of 60 m × 18 m. In the proposed scheme, the total monitoring area is divided into sub-areas by clustering the fingerprint database. The fingerprinting database is constructed using the RSSI measurements from 66 anchor nodes deployed randomly in the given environment. The proposed ANNs are trained for each region by using only those fingerprints that belong to one separate cluster. During online localization of mobile node after off-line training phase, only that fingerprint is utilized, which closely matches with the current RSSI measurement vector. To further refine the obtained location estimates, FCNN is employed. The work in [39] presents an RBF-based localization method to construct RSSI fingerprint for the given area. The difference in RSSI measurements is used as the input to the RBF network to improve the reliability and precision of localization. The advantage of RBF architecture is that it can handle nonlinear estimation problem well and it is easy to train. However, it needs hidden nodes for every applied pattern in training dataset, which generally raises the overall system complexity.

Recently, some research works used particle swarm optimization (PSO) with ANN to solve the target localization problem using wireless networks. In [40], two separate PSO-based approaches are discussed for the localization of sensor nodes. The first proposed approach provides the target locations, whereas the objective of the second approach is to converge the moving nodes around considered target node. The obtained result concludes that PSO based implementation is an efficient collaborative localization and navigation in WSN. This work in [41] is centered around the computation of the distances between the mobile node (here mobile node is bicycle) and anchor node (here it is coach) in the indoor and the outdoor environments. Out of the two proposed approaches, the first approach is based on the traditional LNSM model with the help of the RSSI measurements from Zigbee module, whereas the second is based on a proposed PSO-ANN algorithm to enhance estimation accuracy in distance. The LNSM parameters were estimated using real-time RSSI measurements in both the environments. In the proposed scheme, a feed-forward type of ANN trained with Levenberg-Marquardt (LM) training algorithm is used to locate the mobile node and the coach. The experimental results conclude that the proposed PSO-ANN algorithm significantly reduced the error in distance estimation as compared to that with the traditional LNSM-based approach without the need for additional components. The achieved mean absolute error with the proposed scheme is 0.022 m and 0.208 m for the outdoor and the indoor setup, respectively. Apart from this, the author also investigates the impact of anchor density on localization error for considered indoor setup. The work in [42] presented a feed-forward type of ANN with LM training for position estimation of the mobile node using the RSSI measurements from WSN deployed in the indoor setup. The mobile node collects the RSSI values from five deployed anchor nodes, which are then utilized for training the ANN. The comparison of the proposed method with a weighted k-nearest neighbor (WkNN) method reveals that the localization performance of WkNN is better than that of the proposed ANN-based implementation when five anchor nodes are used. However, if three anchor nodes are considered, then the proposed ANN-based scheme outperforms WkNN. The work in [43] proposed uses Dijkstra and LNSM algorithms to gather the RSSI measurements and to compute the shortest possible distance of the mobile node from the anchor nodes. The simulation is carried out for the area 50 m × 50 m and 80 m × 80 m with a communication range of nodes to 20 m and 25 m, respectively. The simulation results prove the efficacy of the proposed approach as compared to that of PSO and ANN.

The back propagation neural network (BPNN) has also been widely used to indoor L&T problem in optical-, RFID-, WLAN-/Wi-Fi-, and dead reckoning-based systems. In [31], the authors have presented Sigmoidal feed-forward ANN (SFFANN) and RBF based for localization in WSN. The proposed scheme uses the RSSI measurements from three anchor nodes deployed in 100 m × 100 m grid area. The MATLAB simulation results demonstrate that SFFANN-based implementation outperforms as compared to RBF-based implementation in the context of localization accuracy. The mean localization error of 5.15 m is shown by MLP. The proposed SFFANN is trained using BPNN. The relative location of the sensor node is found to be in the proximity of 5.15 m of the actual location in case of SFFANN, whereas in the case of RBF network, it is 6.07 m. The work in [44] proposes to exploit BPNN for both stages of fingerprinting-based indoor localization using WLAN/Wi-Fi RSS. In this implementation, the first stage is the construction radio fingerprint of the given environment during the off-line stage, and the second is the localization of sensor node during the online stage. The authors also show that the proposed BPNN-driven algorithm with only one hidden layer demonstrates networks with better positioning accuracy than kNN and that too with less computational burden during the online stage.

3.4 Survey of Application of BLE Technology in RSSI-Based Target Tracking

Being a cost-effective and low-power technology and having its availability on numerous wireless devices, BLE-based indoor L&T applications are drastically increased in this decade. The BLE-based RSSI measurements have found to be more stable than the Wi-Fi-based RSSI measurements. Easy deployment and low-powered nature make them an ideal candidate for constructing WSN. A number of BLE-based L&T systems have become also popular for LBS.

The existing BLE-based target L&T using the RSSI measurements as a metric can be mainly classified into two approaches: (1) based on trilateration using suitable signal path loss model and (2) based on RF fingerprinting. The important drawback of the former approach is the appropriate choice of the signal path loss model to characterize the given RF environment. Additionally, the fine calibration of parameters of selected signal path loss model to suit with the given environment is a very complex task. In the latter approach, the localization accuracy has been proved to be superior to that with the former approach. However, the major limitation of the latter approach is the need of lot of time for training the underlying system to characterize the considered RF environment. Additionally, RF fingerprinting-based localization system is generally more susceptible to any minute change in the given environment. The RF fingerprinting-based localization system uses radio map of large number of environmental locations to achieve sufficient localization accuracy. The limitations of both of these techniques can be lowered down by coupling them with some suitable modern techniques such KF or PF or their variants.

The research work in [45] presents EKF-based localization scheme, which yields a localization accuracy of 2.56 m with the help of sparsely deployed BLE beacon nodes. The work in [46] presents a three-step cascaded Kalman filter (CKF) to estimate the target position even with environmental dynamicity (i.e., NLOS and multipath) with the help of the BLE and an IMU module. In this work, the target positions are estimated using the fusion of the RSSI measurements from BLE nodes and acceleration measurements from IMU. The estimated locations are further refined using Rauch-Tung-Striebel smoother algorithm. The result confirms that the proposed scheme provides a very low localization error irrespective of a highly dynamic environment. From the research work in [47], the authors proposed a novel mobile application being operated with the help of the communication links of BLE beacon nodes. The first position estimates of moving target are obtained using RF fingerprinting; thereafter, PF algorithm is used to further refine these position estimates to deal with noisy RSSI measurements. The research work in [48] proposes a novel localization algorithm named as "InLoc." Using the proposed algorithm, the initial user locations are estimated using the RSSI measurements obtained from BLE beacons deployed. Instead of using the specially designed vector map, here the InLoc system uses the RF radio map of the building floors. This radio map can also be utilized for PF- and IMU-based routing and tracking. The proposed system yields an average tracking error of less than 0.4 m and an average localization error of 0.9 m. In order to deal with multipath propagation and fluctuations in RSSI; the authors of [49] proposed a real-time indoor L&T system with the help of BLE technology. The three main proposals in the proposed research work are frequency diversity, KF, and a weighted trilateration technique. The experimental results prove that both proposals improve the target tracking performance. The tracking error is found to be around 1.82 m for 90% of the time of target motion in a middle-sized room. The system is found to be efficient and scalable in the context of power consumption and cost.

The important reason for lower localization accuracy in all the above in BLE-based implementations is high and uncertain fluctuations in RSSI measurements. Additionally, due to the use of radio signals with short wavelength, the transfer of data is affected due to factors such as multipath propagation and NLOS. Therefore, still there is a lot of scope to improve the target localization accuracy further in BLE- and RSSI-based implementation. The capability of ANN can be utilized to deal with noisy RSSI measurements as well as issues related with signal propagation such as NLOS and multipath propagation. Very limited research efforts are made on the application of the ANN in the BLE-based indoor L&T. In [50], previously trained ANN by RF fingerprinting is utilized to compute the locations of mobile target using a Bluetooth technique. The RF fingerprinting works in two steps, namely, off-line step and online step. In the off-line step, the designed ANN is trained using the obtained RSSI measurements, whereas in the online step, the proposed system is tested for real-time random input vector of RSSI measurements. The major drawback of the proposed implementation is the requirement to train the ANN with a large database of RSSI and associated 2-D locations, which is highly time-consuming. With the capability of one-pass learning and capability to get trained quickly using very few measurements, the application of probabilistic neural network can be a very economical alternative for target tracking in BLE-based systems [51, 52].

3.5 Limitations in the Existing RSSI-Based L&T Systems

Although a lot of RSSI-based L&T algorithms are proposed in the literature in the past, the increased demand for high tracking accuracy has challenged the research community. As discussed in Sect. 3.3, the major reason for low tracking accuracy in RSSI-based L&T systems is the environmental dynamicity. The two major approaches to solve this problem are as follows: the first is to modify the signal path loss model through finer refinement of model parameters to suit considered RF environment and use a suitable localization technique [19, 53], whereas the second is fusing RSSI measurements with an appropriate Bayesian filter, such as KF and PF. Many RSSI-based target tracking implementations rely on the use of the traditional range-based techniques, such as lateration or angulation. Some RSSI-based implementations also exploit the RF fingerprinting as well as ANN to extract the RSSI-distance relationship for the given environment. The major drawbacks of these existing RSSI-based implementations are the following:

1. The localization error in most of the existing systems in the literature is above 1 m for the indoor environment. However, such low localization accuracy is not sufficient for the indoor L&T applications, which demand a high tracking accuracy. It can be increased by raising anchor density, but it will increase the overall system cost. Although the tracking accuracy of 4.33 cm is achieved in [10], the proposed approach was verified for a very smaller WSN area of 80 cm × 80 cm. The monitoring area in real-time indoor target tracking applications is larger than this in practice.

2. By calibrating the parameters of selected RF path loss model, the target L&T accuracy can be enhanced. However, due to the highly nonlinear RSSI-distance relationship, the proposed systems show severe distance estimation errors, which in turn increase the localization error. Additionally, a slight change in the environmental setup or a completely new environmental setup may not guarantee a high tracking accuracy with the use of previously calibrated parameters of the path loss model. Thus, the proposed algorithms must be robust enough to accommodate the environmental changes.
3. As discussed earlier, majority of RSSI-based L&T systems depend upon the traditional methods, such as angulation or lateration. However due to environmental dynamicity (i.e. noisy RSSI measurements), these techniques suffer from inaccurate computations of distances and angles. Thus, these traditional techniques alone can't offer high tracking accuracy in the context of noisy RSSI measurements. Therefore, some advanced techniques must be coupled with them to enhance the tracking performance.
4. Many Bayesian framework-based implementations for target L&T can be found in the literature. Although this approach can provide the tracking accuracy around 1 m, however, very few researchers mentioned the computational time required for the execution of proposed algorithms.
5. In practice, the target moves with mobility pattern such as constant velocity (CV) or constant acceleration motion (CA) and the abrupt variations in the target velocity. None of the research works reported so far presented the investigation of their developed approach for the impact of variation in the target mobility model in the context of environmental dynamicity.
6. Many RSSI-based implementations in the literature used RF fingerprinting to characterize the given wireless environment. Although there is no need to compute distances for target L&T in this approach, creating an RF fingerprint offline is a very complex as well as time consuming task. Additionally, the researchers have not investigated the target tracking performance for the sudden changes in the environment.
7. Although few ANN architectures have been adopted by the researchers in the past for indoor RSSI-based target tracking, the training of these architectures is very time-consuming. Also, these ANN architectures require more iteration to learn the dynamicity of the given environment.
8. The realization and validation of proposed RSSI-based target tracking algorithm on the actual WSN hardware (test bed) is a highly challenging task. Many existing works in the literature do not attempt to validate the proposed algorithm on the actual WSN hardware setup.

The above drawbacks in the existing RSSI-based target tracking systems highlight that there is still a lot of scope to enhance the tracking performance further in the context of high environmental dynamicity, limited RSSI measurements, and abrupt variations in target velocity. In order to get real-time performance, the computational complexity of developed algorithm must be as small as possible. These important research challenges are to be addressed by the researchers of this domain.

References

1. H.D. Chon, S. Jun, H. Jung, W. An, Using RFID for accurate positioning. J. Glob. Position. Syst. (2010). https://doi.org/10.5081/jgps.3.1.32
2. F. Viani, P. Rocca, G. Oliveri, D. Trinchero, A. Massa, Localization, tracking, and imaging of targets in wireless sensor networks: an invited review. Radio Sci. (2011). https://doi.org/10.1029/2010RS004561
3. Y. Zhang, X. Li, M. Amin, Principles and techniques of RFID positioning, in *RFID Systems: Research Trends and Challenges*, (Wiley, New York, 2010)
4. J. Zhu, H. Xu, Review of RFID-based indoor positioning technology, in *Innovative Mobile and Internet Services in Ubiquitous Computing. IMIS 2018. Advances in Intelligent Systems and Computing*, (Springer, Cham, 2019). https://doi.org/10.1007/978-3-319-93554-6_62
5. M.R. Mahfouz, A.E. Fathy, M.J. Kuhn, Y. Wang, Recent trends and advances in UWB positioning, in *Innovative Mobile and Internet Services in Ubiquitous Computing. IMIS 2018. Advances in Intelligent Systems and Computing*, (Springer, Cham, 2009). https://doi.org/10.1109/IMWS2.2009.5307895
6. X. Zhao, Z. Xiao, A. Markham, N. Trigoni, Y. Ren, Does BTLE measure up against WiFi? A comparison of indoor location performance, in *European Wireless 2014; 20th European Wireless Conference* (2014)
7. R. Faragher, R. Harle, An analysis of the accuracy of bluetooth low energy for indoor positioning applications, in *International Technical Meeting* of the *Satellite Division of the Institute of Navigation* (2014). https://doi.org/10.1016/j.ajpath.2011.02.025
8. R. Faragher, R. Harle, Location fingerprinting with bluetooth low energy beacons. IEEE J. Sel. Areas Commun. (2015). https://doi.org/10.1109/JSAC.2015.2430281
9. R. Mardeni, S. Nizam, Node positioning in zigbee network using trilateration method based on the received signal strength indicator (RSSI). Eur. J. Sci. Res. **46**, 48–61 (2010)
10. G. Blumrosen, T. Anker, B. Hod, D. Dolev, B. Rubinsky, Enhancing RSSI-based tracking accuracy in wireless sensor networks. ACM Trans. Sens. Netw. (2013). https://doi.org/10.1145/2480730.2480732
11. F. Gustafsson et al., Particle filters for positioning, navigation, and tracking. IEEE Trans. Signal Process. (2002). https://doi.org/10.1109/78.978396
12. D. Fox, J. Hightower, L. Liao, D. Schulz, G. Bordello, Bayesian filtering for location estimation. IEEE Pervasive Comput., (2003). https://doi.org/10.1109/MPRV.2003.1228524
13. L. Mihaylova, D. Angelova, D. R. Bull, N. Canagarajah, Localization of mobile nodes in wireless networks with correlated in time measurement noise. IEEE Trans. Mob. Comput. (2011). https://doi.org/10.1109/TMC.2010.132
14. J. Hightower, G. Borriello, Location systems for ubiquitous computing. Computer (2001). https://doi.org/10.1109/2.940014
15. J. Letchner, D. Fox, A. LaMarca, Large-scale localization from wireless signal strength, in *Association for the Advancement of Artificial Intelligence* (2005)
16. A. R. Destiarti, P. Kristalina, A. Sudarsono, Mobile cooperative tracking with RSSI ranging in EKF algorithm for indoor wireless sensor network, in *2016 International Conference on Knowledge Creation and Intelligent Computing (KCIC)* (2017). https://doi.org/10.1109/KCIC.2016.7883626
17. N. B. Priyantha, A. Chakraborty, H. Balakrishnan, The Cricket location-support system, in *Proceedings of the 6th Annual International Conference on Mobile Computing and Networking* (2003). https://doi.org/10.1145/345910.345917
18. F. Zafari, I. Papapanagiotou, T. J. Hacker, A novel bayesian filtering based algorithm for RSSI-based indoor localization, in *2018 IEEE International Conference on Communications (ICC)* (2018). https://doi.org/10.1109/ICC.2018.8423012
19. F. Zafari, I. Papapanagiotou, M. Devetsikiotis, T. J. Hacker, Enhancing the accuracy of iBeacons for indoor proximity-based services, in *2017 IEEE International Conference on Communications (ICC)* (2017). https://doi.org/10.1109/ICC.2017.7996508

20. Z. Chen, H. Zou, H. Jiang, Q. Zhu, Y. C. Soh, L. Xie, Fusion of WiFi, smartphone sensors and landmarks using the kalman filter for indoor localization. Sensors (Switzerland) (2015). https://doi.org/10.3390/s150100715
21. Z. Chen, Q. Zhu, Y. C. Soh, Smartphone inertial sensor-based indoor localization and tracking with iBeacon corrections. IEEE Trans. Ind. Inf. (2016). https://doi.org/10.1109/TII.2016.2579265
22. J. Röbesaat, P. Zhang, M. Abdelaal, O. Theel, An improved BLE indoor localization with Kalman-based fusion: an experimental study. Sensors (Switzerland) (2017). https://doi.org/10.3390/s17050951
23. M. Ben Kilani, A. J. Raymond, F. Gagnon, G. Gagnon, P. Lavoie, RSSI-based indoor tracking using the extended Kalman filter and circularly polarized antennas, in *2014 11th Workshop on Positioning, Navigation and Communication (WPNC)* (2014). https://doi.org/10.1109/WPNC.2014.6843305
24. B. Labinghisa, G. S. Park, D. M. Lee, Improved indoor localization system based on virtual access points in a Wi-Fi environment by filtering schemes, in *2017 International Conference on Indoor Positioning and Indoor Navigation (IPIN)* (2017). https://doi.org/10.1109/IPIN.2017.8115873
25. A. S. Paul, E. A. Wan, RSSI-Based indoor localization and tracking using sigma-point kalman smoothers. IEEE J. Sel. Top. Signal Process. (2009). https://doi.org/10.1109/JSTSP.2009.2032309
26. X. Peng, K. Yang, C. Liu, Maneuvering target tracking using current statistical model based adaptive UKF for wireless sensor network. J. Commun. (2015). https://doi.org/10.12720/jcm.10.8.579-588
27. V. Malyavej, P. Udomthanatheera, RSSI/IMU sensor fusion-based localization using unscented Kalman filter, in *The 20th Asia-Pacific Conference on Communication (APCC2014)* (2015). https://doi.org/10.1109/APCC.2014.7091638
28. M. Gholami, N. Cai, R. W. Brennan, An artificial neural network approach to the problem of wireless sensors network localization. Robot. Comput. Integr. Manuf. (2013). https://doi.org/10.1016/j.rcim.2012.07.006
29. A. Shareef, Y. Zhu, M. Musavi, Localization using neural networks in wireless sensor networks, in *Proceedings of the 1st International Conference on MOBILe Wireless MiddleWARE, Operating Systems, and Applications* (2009). https://doi.org/10.4108/icst.mobilware2008.2901
30. S. H. Chagas, J. B. Martins, L. L. De Oliveira, An approach to localization scheme of wireless sensor networks based on artificial neural networks and genetic algorithms, in *10th IEEE International NEWCAS Conference* (2012). https://doi.org/10.1109/NEWCAS.2012.6328975
31. A. Payal, C. S. Rai, B. V. R. Reddy, Analysis of some feedforward artificial neural network training algorithms for developing localization framework in wireless sensor networks. Wirel. Pers. Commun. (2015). https://doi.org/10.1007/s11277-015-2362-x
32. S. Yun, J. Lee, W. Chung, E. Kim, S. Kim, A soft computing approach to localization in wireless sensor networks. Expert Syst. Appl. (2009). https://doi.org/10.1016/j.eswa.2008.09.064
33. N. Irfan, M. Bolic, M. C. E. Yagoub, V. Narasimhan, Neural-based approach for localization of sensors in indoor environment. Telecommun. Syst. (2010). https://doi.org/10.1007/s11235-009-9223-4
34. M. Stella, M. Russo, D. Begusic, Location determination in indoor environment based on RSS fingerprinting and artificial neural network, in *2007 9th International Conference on Telecommunications* (2007). https://doi.org/10.1109/CONTEL.2007.381886
35. T. Ogawa, S. Yoshino, M. Shimizu, H. Suda, A new in-door location detection method adopting learning algorithms, in *Proceedings of the First IEEE International Conference on Pervasive Computing and Communications, 2003. (PerCom 2003)* 2004. https://doi.org/10.1109/percom.2003.1192784
36. S. Mahfouz, F. Mourad-Chehade, P. Honeine, J. Farah, H. Snoussi, Target tracking using machine learning and kalman filter in wireless sensor networks. IEEE Sensors J. (2014). https://doi.org/10.1109/JSEN.2014.2332098

37. S. Kumar, S. Lee, Localization with RSSI values for wireless sensor networks: an artificial neural network approach, in *International Electronic Conference on Sensors and Applications* (2014). https://doi.org/10.3390/ecsa-1-d007
38. M. Bernas, B. Płaczek, Fully connected neural networks ensemble with signal strength clustering for indoor localization in wireless sensor networks. Int. J. Distrib. Sens. Netw. (2015). https://doi.org/10.1155/2015/403242
39. D. Guo, Y. Zhang, Q. Xiang, Z. Li, Improved radio frequency identification indoor localization method via radial basis function neural network. Math. Probl. Eng. (2014). https://doi.org/10.1155/2014/420482
40. R. V. Kulkarni, G. K. Venayagamoorthy, A. Miller, C. H. Dagli, Network-centric localization in MANETs based on particle swarm optimization, in *2008 IEEE Swarm Intelligence Symposium* (2008). https://doi.org/10.1109/SIS.2008.4668299
41. S. K. Gharghan, R. Nordin, M. Ismail, J. A. Ali, Accurate wireless sensor localization technique based on hybrid PSO-ANN algorithm for indoor and outdoor track cycling. IEEE Sensors J. (2016). https://doi.org/10.1109/JSEN.2015.2483745
42. L. Gogolak, S. Pletl, D. Kukolj, Neural network-based indoor localization in WSN environments. Acta Polytech. Hungarica **10**, 221–235 (2013)
43. P.-J. Chuang, Y.-J. Jiang, Effective neural network-based node localisation scheme for wireless sensor networks. IET Wirel. Sens. Syst. (2014). https://doi.org/10.1049/iet-wss.2013.0055
44. C. Zhou, A. Wieser, Application of backpropagation neural networks to both stages of fingerprinting based WIPS, in *2016 Fourth International Conference on Ubiquitous Positioning, Indoor Navigation and Location Based Services (UPINLBS)* (2017). https://doi.org/10.1109/UPINLBS.2016.7809973
45. Y. Zhuang, J. Yang, Y. Li, L. Qi, N. El-Sheimy, Smartphone-based indoor localization with bluetooth low energy beacons. Sensors (Switzerland) (2016). https://doi.org/10.3390/s16050596
46. P. K. Yoon, S. Zihajehzadeh, B. S. Kang, E. J. Park, Adaptive Kalman filter for indoor localization using bluetooth low energy and inertial measurement unit, in *2015 37th Annual International Conference of the IEEE Engineering in Medicine and Biology Society (EMBC)* (2015). https://doi.org/10.1109/EMBC.2015.7318489
47. Y. Inoue, A. Sashima, K. Kurumatani, Indoor positioning system using beacon devices for practical pedestrian navigation on mobile phone, in *Ubiquitous Intelligence and Computing*, (2009). https://doi.org/10.1007/978-3-642-02830-4_20
48. V. Chandel, N. Ahmed, S. Arora, A. Ghose, Inloc: an end-to-end robust indoor localization and routing solution using mobile phones and ble beacons, in *2016 International Conference on Indoor Positioning and Indoor Navigation (IPIN)* (2016). https://doi.org/10.1109/IPIN.2016.7743592
49. V. C. Paterna, A. C. Augé, J. P. Aspas, M. A. P. Bullones, A bluetooth low energy indoor positioning system with channel diversity, weighted trilateration and kalman filtering. Sensors (Switzerland) (2017). https://doi.org/10.3390/s17122927
50. M. Altini, D. Brunelli, E. Farella, L. Benini, Bluetooth indoor localization with multiple neural networks, in *IEEE 5th International Symposium on Wireless Pervasive Computing* (2010). https://doi.org/10.1109/ISWPC.2010.5483748
51. D. F. Specht, Probabilistic neural networks. Neural Netw. (1990). https://doi.org/10.1016/0893-6080(90)90049-Q
52. D.F. Specht, A general regression neural network. IEEE Trans. Neural Netw. (1991). https://doi.org/10.1109/72.97934
53. Y. Miao, H. Wu, L. Zhang, The accurate location estimation of sensor node using received signal strength measurements in large-scale farmland. J. Sensors (2018). https://doi.org/10.1155/2018/2325863

Chapter 4
Trilateration-Based Target L&T Using RSSI

4.1 System Assumptions and Design for Trilateration-Based L&T

The proposed system contains a set of stationary anchor nodes at some fixed locations. These anchor nodes are supposed to be placed in 100 m by 100 m area, whereas a single target moves through this area, as shown in Figs. 4.3, 4.5, 4.7, and 4.9. The base station is assumed to be located outside the WSN area. The proposed L&T system is supposed to run for the total period of T, which are divided into many time slots, each of which is denoted by notation dt. The mobile target is supposed to have a WSN node, and it broadcasts RF signal to all the anchor nodes on a time instance k. That means, it is assumed that the anchor nodes act as receivers, and the target is a transmitter in the proposed experimental setup. It can be considered as a case of cooperative L&T. Each anchor node computes their distances from the moving target using received RSSI measurements. The detailed mathematics behind these computations is explained with the equations given in this section. The anchor nodes send these distances and their coordinates to the base station, kept outside the WSN area considered for the proposed system. The base station then selects the lowest three distances out of them. The base station is assumed to be a laptop (Processor: Core i3, 1.89 GHz, 2 GB RAM). Upon receiving the details from anchor nodes, the base station runs the trilateration algorithm to estimate the moving target location for each of the sampling interval. The sensor node communication range is considered 100 m during simulation experiments. The transmission power is assumed to be 1 mW, whereas the receiver and transmitter antenna gains are 1 dBm.

The target is assumed to go through the abrupt variations in the velocity during T sec as per Eqs. (4.1)–(4.5). These variations in target velocity are depicted in Figs. 4.1 and 4.2. The negative value for target velocity means that the target motion

S. R. Jondhale et al., *Received Signal Strength Based Target Localization and Tracking Using Wireless Sensor Networks*, EAI/Springer Innovations in Communication and Computing, https://doi.org/10.1007/978-3-030-74061-0_4

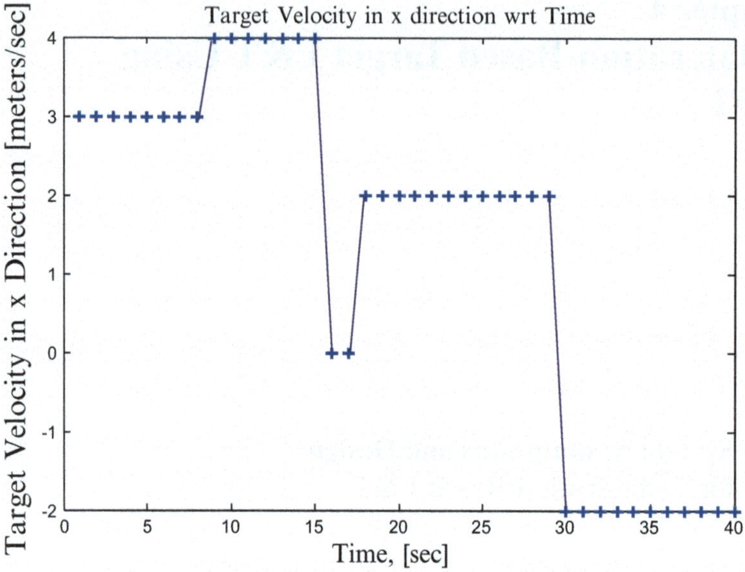

Fig. 4.1 Abrupt variation in target velocity in X-direction during motion

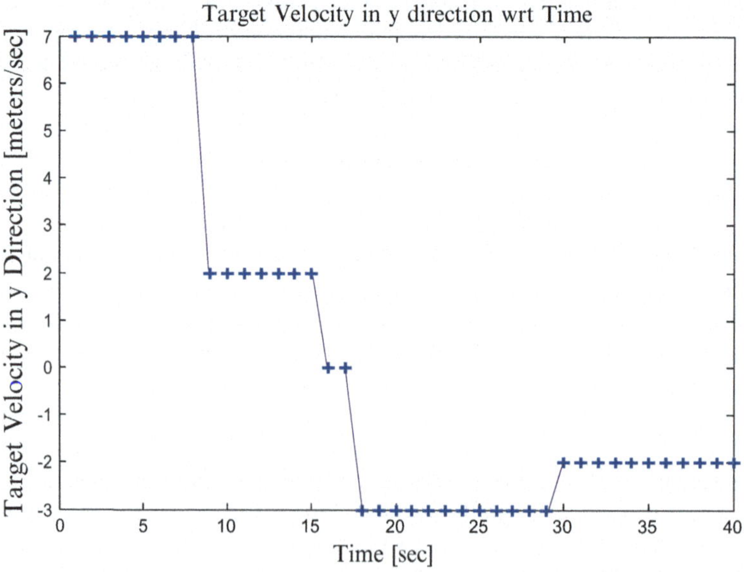

Fig. 4.2 Abrupt variation in target velocity in Y-direction during motion

is toward a location of small coordinate value as compared to the location for previous instance of time.

$$\dot{x}_k = 3, \quad \dot{y}_k = 7, \qquad \text{for } 0 < k < 9 \text{ s},\tag{4.1}$$

$$\dot{x}_k = 4, \quad \dot{y}_k = 2, \qquad \text{for } 9 \le k \le 16 \text{ s},\tag{4.2}$$

$$\dot{x}_k = 0, \quad \dot{y}_k = 0, \qquad \text{for } 16 \le k \le 18 \text{ s},\tag{4.3}$$

$$\dot{x}_k = 2, \quad \dot{y}_k = -3, \qquad \text{for } 18 \le k \le 30 \text{ s}.\tag{4.4}$$

$$\dot{x}_k = -2, \quad \dot{y}_k = -2, \qquad \text{for } 30 \le k \le 40 \text{ s}.\tag{4.5}$$

This chapter follows LNSM model for simulation experiments. The RSSI ($z_{lj,k}$) at the node N_l with coordinates (x_{lk}, y_{lk}) at time instance k, when transmitted from the node N_j with coordinates (x_{jk}, y_{jk}), is based on Eq. (4.6) as given below [1–3]:

$$z_{lj,k} = P_r\left(d_0\right) - 10n\log\left(d_{lj,k} / d_0\right) + X_\sigma,\tag{4.6}$$

As we know the nature or behavior of wireless channels is generally different because of variations in the obstacles between the receiver and the transmitter; hence, the values of n and $P_r(d_0)$ need to be chosen carefully [4–7]. To address environmental dynamicity in RF environment, average values of n (n_{avg}) and $P_r(d_0)$ are computed in advance during system calibration step [See Eqs. (4.7)–(4.12)]. Given three distances (d_1, d_2 and d_3), and three RSSI measurements (z_3, z_2 and z_3), the value of n (n_{avg}) is calculated as given below:

$$z_1 = P_r\left(d_0\right) - 10n_1\log\left(d_1 / d_0\right) + X_\sigma,\tag{4.7}$$

$$z_2 = P_r\left(d_0\right) - 10n_2\log\left(d_2 / d_0\right) + X_\sigma,\tag{4.8}$$

$$z_3 = P_r\left(d_0\right) - 10n_3\log\left(d_3 / d_0\right) + X_\sigma.\tag{4.9}$$

where (n_1, n_2 and n_3) are path loss exponents, corresponding to distances d_1, d_2 and d_3, respectively. By rearranging and subtracting the above equations, the values of n_1, n_2 and n_3 can be easily computed. Thereafter, the average path loss exponent n_{avg} can be calculated by taking average of n_1, n_2 and n_3 as given below:

$$n_{\text{avg}} = \left(n_1 + n_2 + n_3\right)/3\tag{4.10}$$

Therefore, Eq. (4.6) is modified as:

$$z_{lj,k} = P_r\left(d_0\right) - 10n_{\text{avg}}\log\left(d_{lj,k} / d_0\right) + X_\sigma.\tag{4.11}$$

The value of $P_r(d_0)$ can then be calculated using Eq. (4.11) by utilizing the value of RSSI $z_{lj,k}$ for a given distance $d_{lj,k}$ and value of n_{avg}.

$$P_r\left(d_0\right) = z_{lj,k} + 10n_{\text{avg}}\log\left(d_{lj,k} / d_0\right) - X_\sigma.\tag{4.12}$$

The distance between the receiver and transmitter can be computed as given below:

$$d_{lj,k} = d_0 10^{\left(P_r(d_0) - z_{lj,k} + X_\sigma\right) / 10 n_{\text{avg}}}. \tag{4.13}$$

In the carried out research work, two cases have been investigated:

- *Case I: Testing the impact of environmental dynamicity on L&T*
- The objective is to test the efficacy of the trilateration algorithm with four anchors by varying noise in RSSI measurements from 0 dBm to 5 dBm.
- *Case II: Testing the impact of anchor density on L&T*
- The objective is to test the impact of anchor density on the trilateration-based L&T algorithm with 4, 6, and 8 anchors (measurement noise is set to 3 dBm).

4.2 Flow of Trilateration-Based L&T Algorithm

The simulation during one time instance k contains three steps. The first step is the environmental calibration (off-line step) (computation of n_{avg} and $P_r(d_0)$ empirically). The second part is the computation of distances, whereas the third part is the application of an input (three smallest distances of the anchor nodes from the moving target (obtained in the second part) and their coordinates) to the trilateration algorithm. The obtained simulation results are based on the average of more than 50 trials of the proposed trilateration algorithm. The detailed flow of the trilateration-based L&T algorithm during time instance k is described below in Table 4.1:

Table 4.1 Trilateration-based target L&T algorithm

I. Environmental calibration
Step 1: Compute n_{avg} and $P_r(d_0)$.
II. At $k = 0$ s
Step 2: Anchor nodes receive RSSI measurements for every kth instance of time from the target, which are to be utilized to compute their distances ($d_1, d_2, \ldots d_n$) from the mobile target.
Step 3: Anchors transmit the calculated distances ($d_1, d_2, \ldots d_n$) and the corresponding coordinates to the base station. The base station then selects the three smallest distances and the corresponding coordinates of anchor nodes.
III. Localization at the base station
Step 4: The base station then runs the trilateration steps to estimate the x and y coordinates of the target using step 3.
For $k = 1, 2, 3, \ldots T$ s
Step 5: Steps 1–4 are then repeated for above k time steps up to time instance T.
Step 6: Calculate the average localization error and RMSE from the estimated track of target.

4.3 Performance Metrics for Assessment of L&T Performance

To assess the localization performance of the trilateration algorithm, two performance metrics are used, namely, RMSE and average localization error [8–10]. These metrics represent the average estimation error in the target location $(\hat{x}_k,\hat{y}_k)$ and how much close the estimated target location $(\hat{x}_k,\hat{y}_k)$ and actual target location (x_k, y_k), respectively. These two metrics collectively represent a measure of target L&T accuracy. The lower the values of these parameters, the higher the target L&T accuracy would be. For each of the time instance k, the error in x estimate $(\hat{x}_k - x_k)$ and error in y estimate $(\hat{y}_k - y_k)$ for trilateration-based implementation is computed for both Case I and Case II. For each simulation run, the RMSE and average localization error are computed with the help of Eqs. (4.14) and (4.15), respectively.

1. *Average Localization Error*

$$\text{Average Localization Error} = \frac{1}{T}\sum_{k=1}^{T}\frac{(\hat{x}_k - x_k)+(\hat{y}_k - y_k)}{2}, \tag{4.14}$$

2. *Root Mean Square Error (RMSE)*

$$\text{RMSE} = \sqrt{\frac{1}{T}\sum_{k=1}^{T}\frac{(\hat{x}_k - x_k)^2+(\hat{y}_k - y_k)^2}{2}}. \tag{4.15}$$

4.4 Discussion on Results

In Case I, the anchor nodes are stationed at locations as given in Table 4.2.

In both Case I and Case II as mentioned earlier, the target starts moving from location (13, 18) and stops at (64, 22) as shown in Figs. 4.3, 4.5, 4.7, and 4.9. These figures depict the actual and the estimated target trajectories by the trilateration algorithm. The color combinations for various terms in these diagrams are as given below:

For anchor nodes: black filled squares.
For actual target location at any time instance k: red unfilled squares.

Table 4.2 Anchor node locations in deployed WSN

Anchor node	2-D location (in meters)
1	(0, 0)
2	(100, 0)
3	(0, 100)
4	(100, 100)

For estimated location using trilateration: black plus symbol.

The initial state vector of the mobile target is [12, 15, 0, 0]. The target state vector at time instance k is defined as $X_k = (x_k, y_k, \dot{x}_k, \dot{y}_k)$. In this chapter, the target motion is defined using Eqs. (4.16) and (4.17):

$$x_k = x_{k-1} + \dot{x}_k \; dt , \tag{4.16}$$

$$y_k = y_{k-1} + \dot{y}_k \; dt , \tag{4.17}$$

4.4.1 *Case I Results: Testing the Impact of Environmental Dynamicity on L&T (Variation in RSSI Measurement Noise)*

The extensive simulation experiments carried out demonstrate high target tracking accuracy, irrespective of the environmental dynamicity and abrupt variations in target velocity. There are many parameters that impact the performance of RSSI-based L&T algorithm, namely, changes in the target velocity, anchor density, and measurement noise in the given RF environment. The higher the anchor density, the higher the target tracking accuracy would be. To study this effect, during simulation experimentation the anchor density is varied from 4 to 8 in steps of 2. To understand the effect of abrupt variations in target velocity, we varied velocity abruptly in the range of −2 to 7 m/s at specific time instances. Whereas to study the impact of environmental dynamicity (variations in fluctuations), the measurement noise in the RSSI is changed from 0 dBm to 5 dBm during Case I (Table 4.3, Figs. 4.3, 4.4, 4.5, 4.6, 4.7, 4.8, 4.9, 4.10, 4.11, 4.12, 4.13, 4.14, 4.15, 4.16, 4.17, 4.18, 4.19, 4.20, 4.21, 4.22, 4.23, 4.24, 4.25, and 4.26).

Table 4.3 Numeric results of Case I: target L&T performance

Four anchor case			
Sr no	Measurement noise (in dBm)	Average localization error (in meters)	RMSE
1	0	3.55	6.99
2	1	4.01	7.03
3	2	4.66	8.43
4	3	6.87	11.02
5	4	8.61	13.70
6	5	9.68	15.55

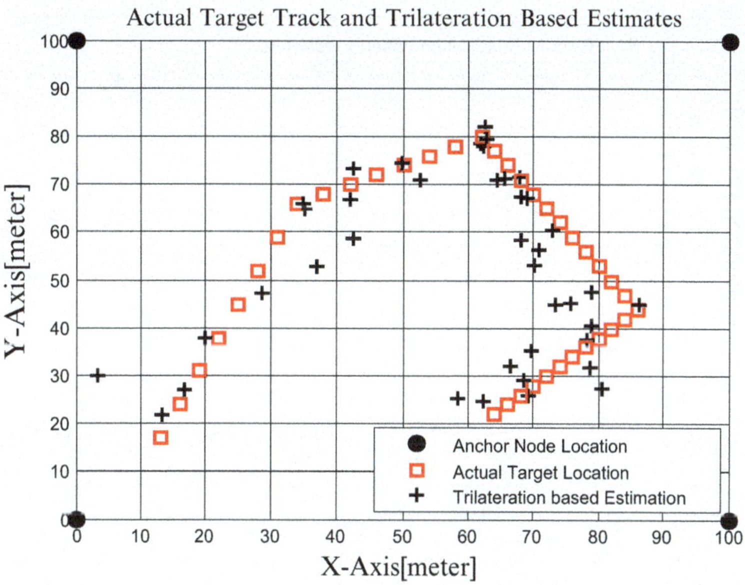

Fig. 4.3 Actual and estimated target trajectories using trilateration algorithm (measurement noise = 0 dBm)

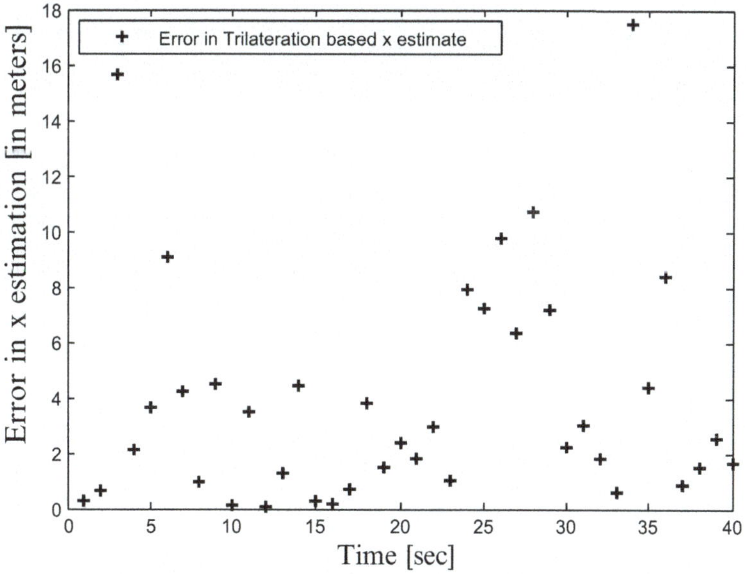

Fig. 4.4 Localization error in *x* estimate (measurement noise = 0 dBm)

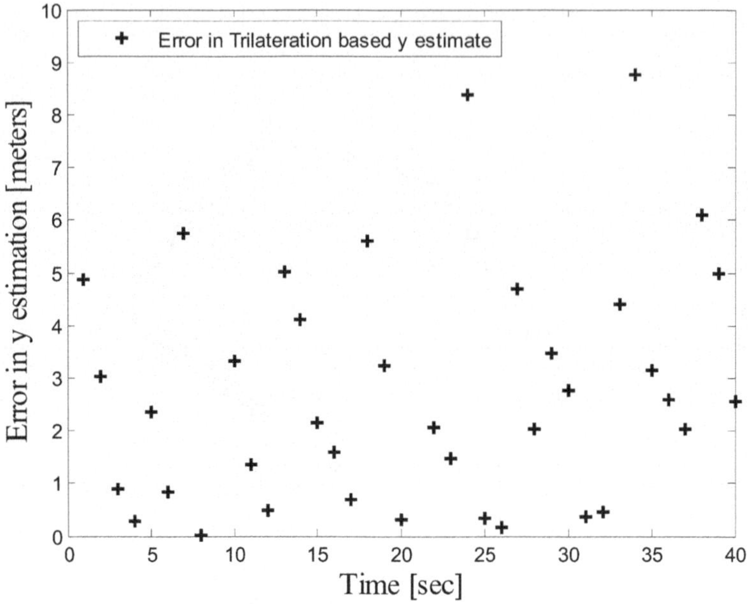

Fig. 4.5 Localization error in y estimate (measurement noise = 0 dBm)

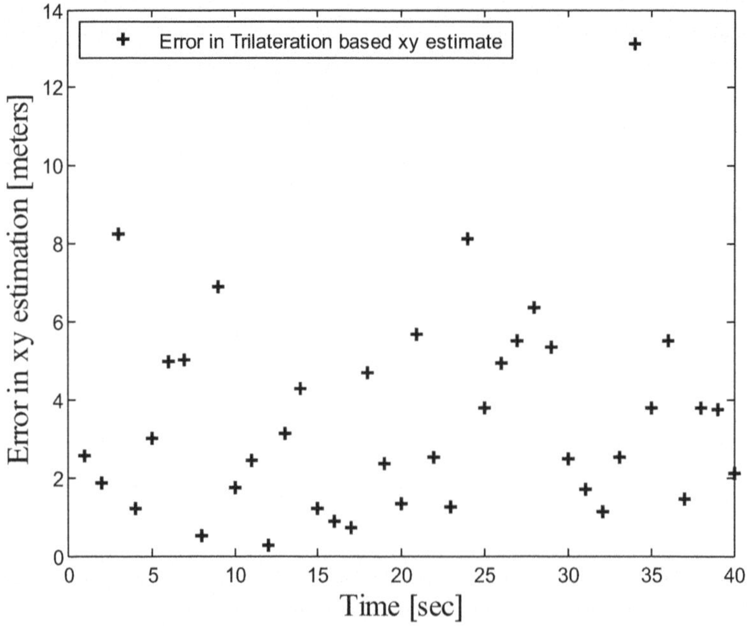

Fig. 4.6 Localization error in *x–y* estimate (measurement noise = 0 dBm)

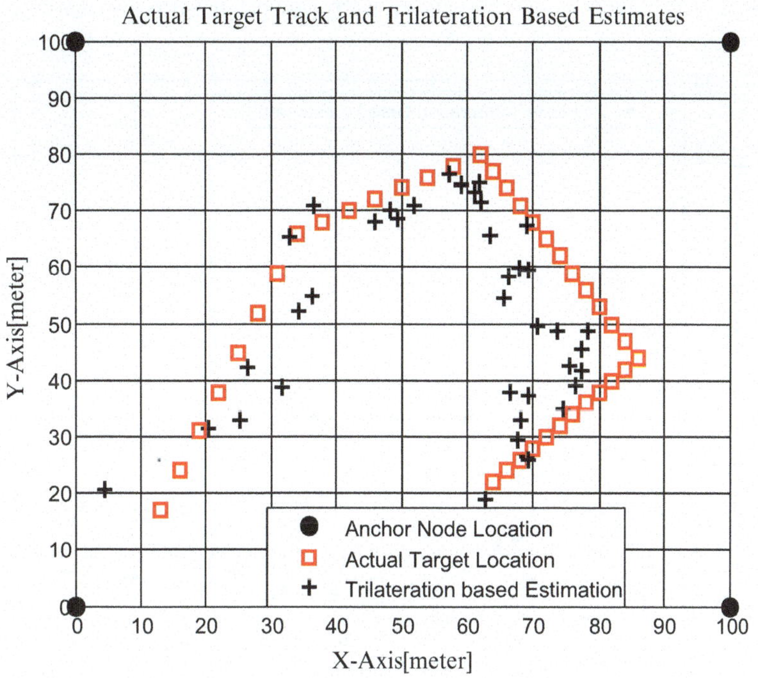

Fig. 4.7 Actual and estimated target trajectories using trilateration algorithm (measurement noise = 1 dBm)

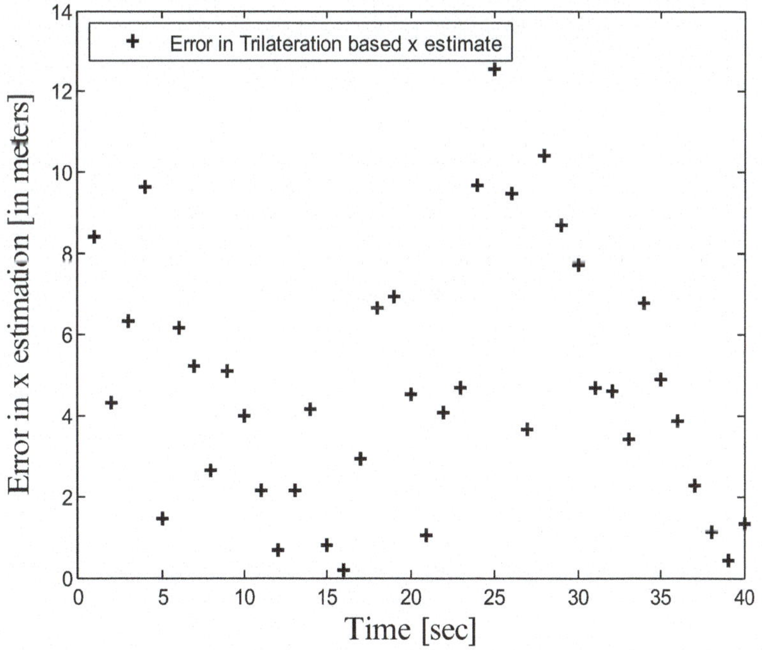

Fig. 4.8 Localization error in *x* estimate (measurement noise = 1 dBm)

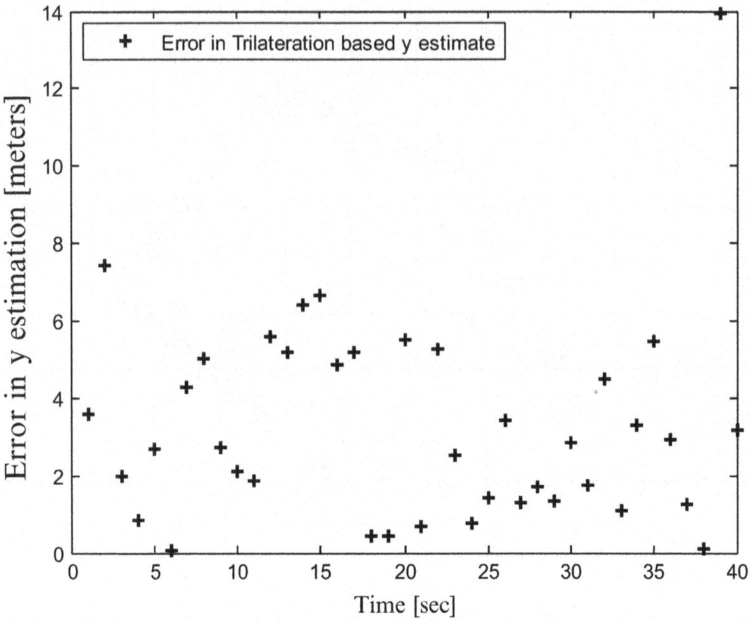

Fig. 4.9 Localization error in *y* estimate (measurement noise = 1 dBm)

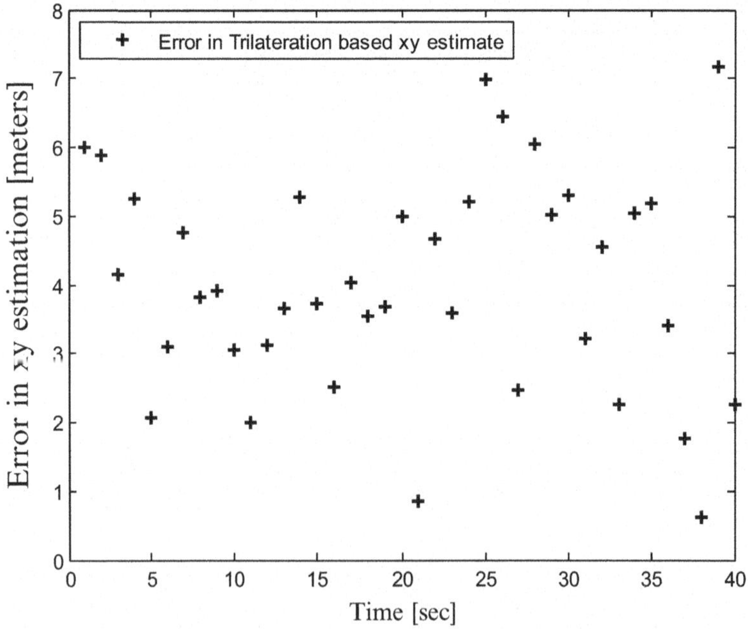

Fig. 4.10 Localization error in *x*–*y* estimate (measurement noise = 1 dBm)

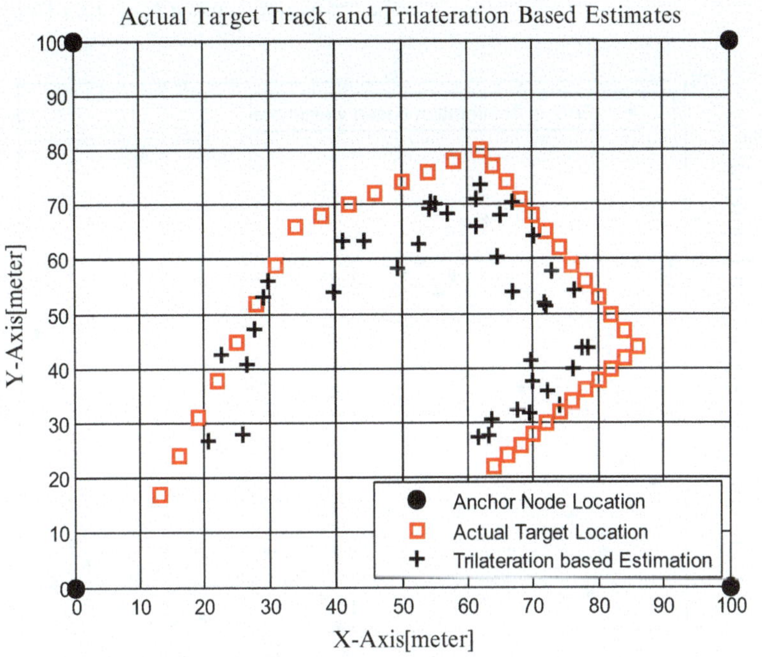

Fig. 4.11 Actual and estimated target trajectories using trilateration algorithm (measurement noise = 2 dBm)

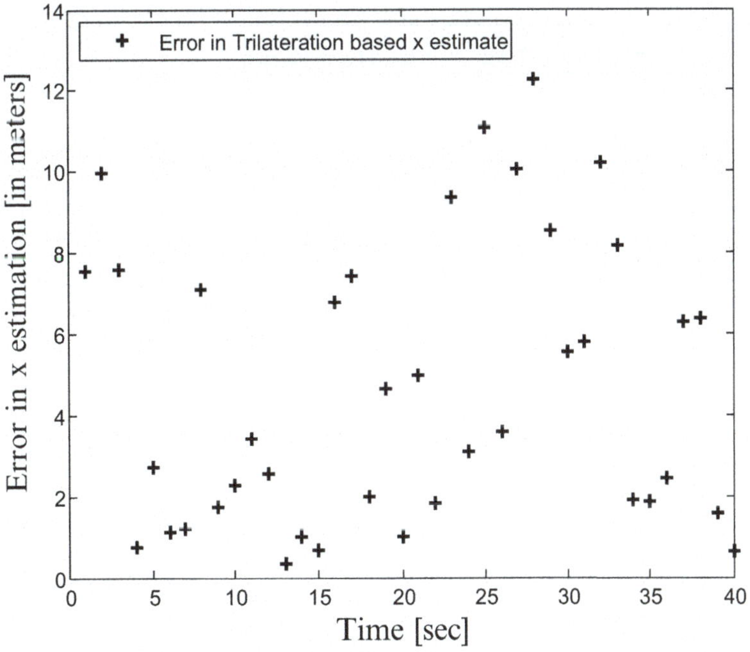

Fig. 4.12 Localization error in *x* estimate (measurement noise = 2 dBm)

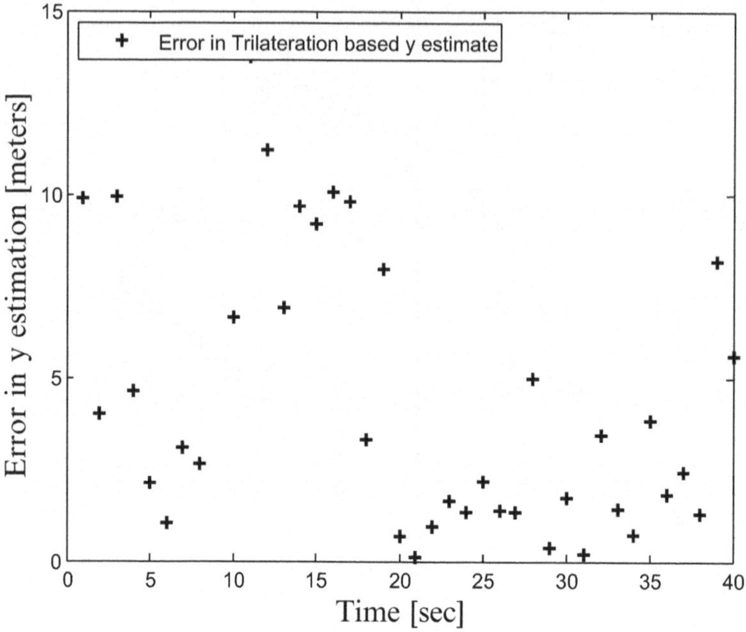

Fig. 4.13 Localization error in y estimate (measurement noise = 2 dBm)

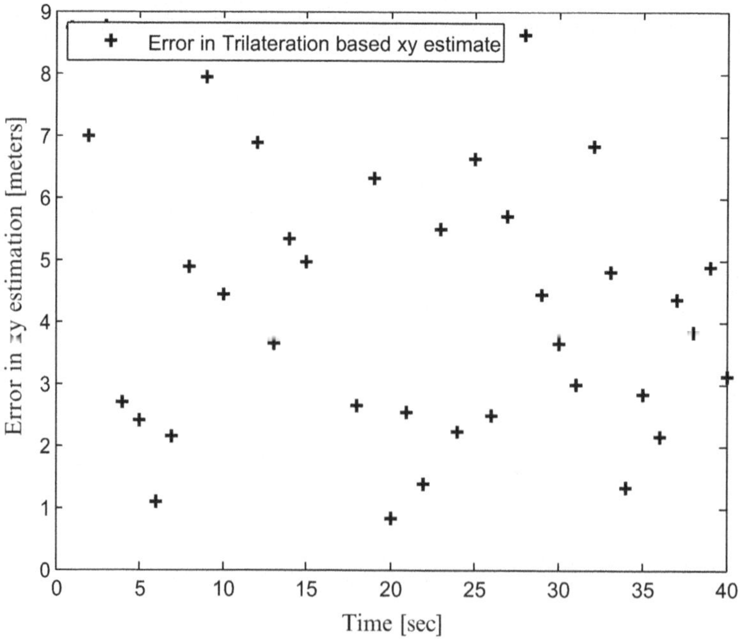

Fig. 4.14 Localization error in x–y estimate (measurement noise = 2 dBm)

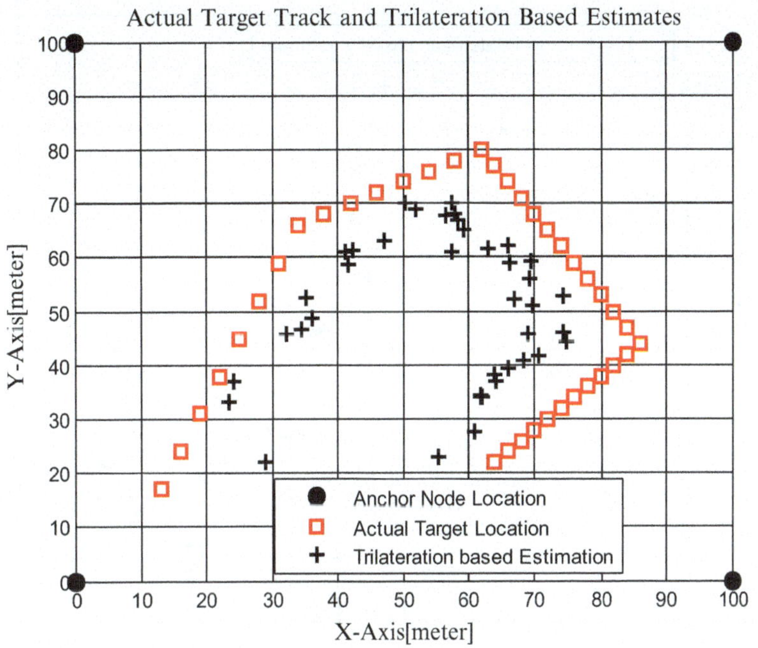

Fig. 4.15 Actual and estimated target trajectories using trilateration algorithm (measurement noise = 3 dBm)

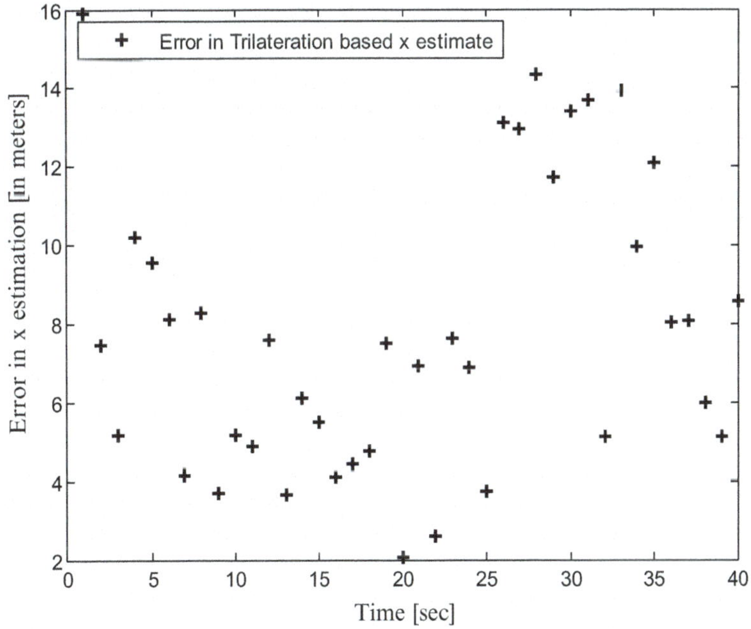

Fig. 4.16 Localization error in *x* estimate (measurement noise = 3 dBm)

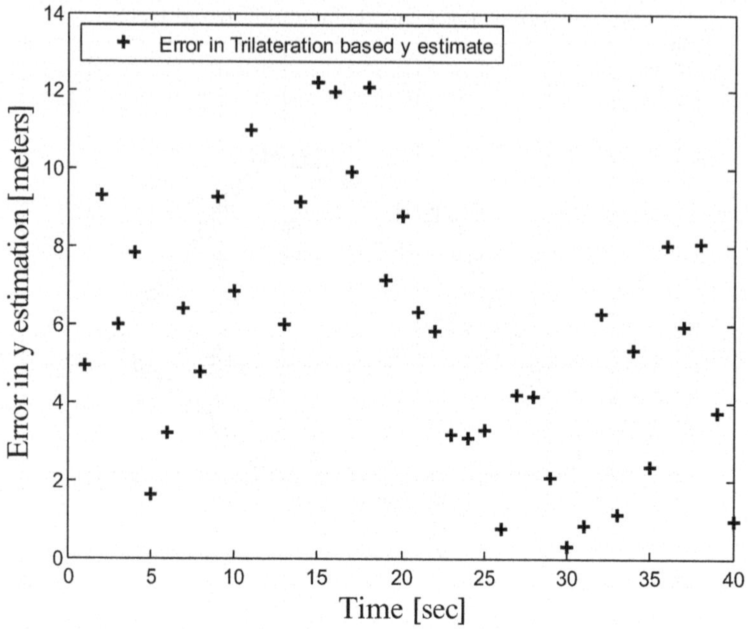

Fig. 4.17 Localization error in *y* estimate (measurement noise = 3 dBm)

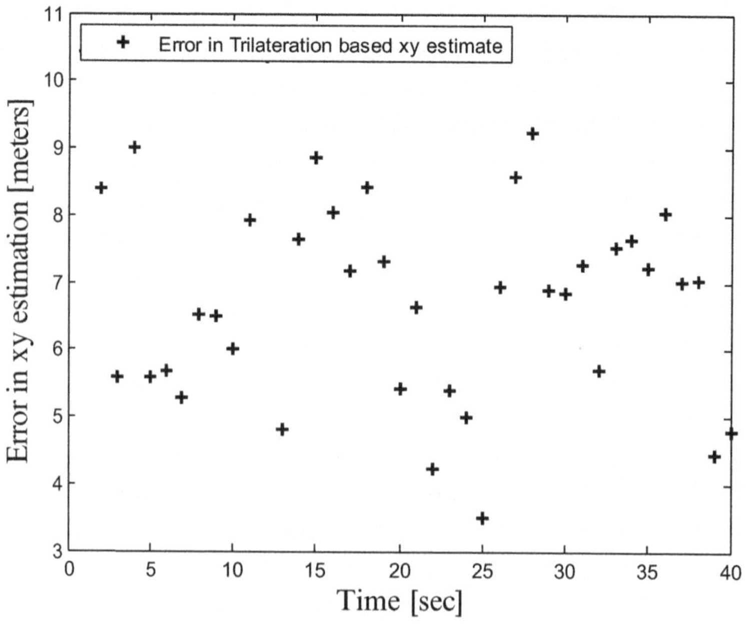

Fig. 4.18 Localization error in *y* estimate (measurement noise = 3 dBm)

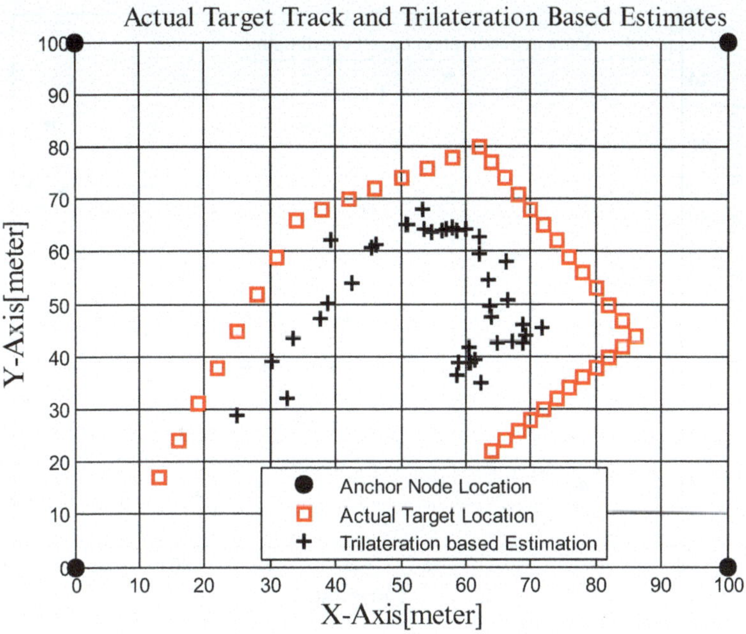

Fig. 4.19 Actual and estimated target trajectories using trilateration algorithm (measurement noise = 4 dBm)

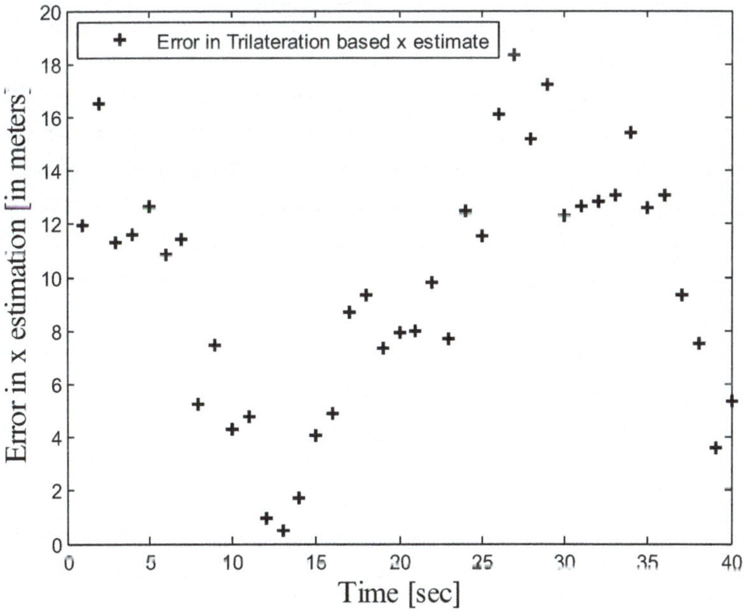

Fig. 4.20 Localization error in x estimate (measurement noise = 4 dBm)

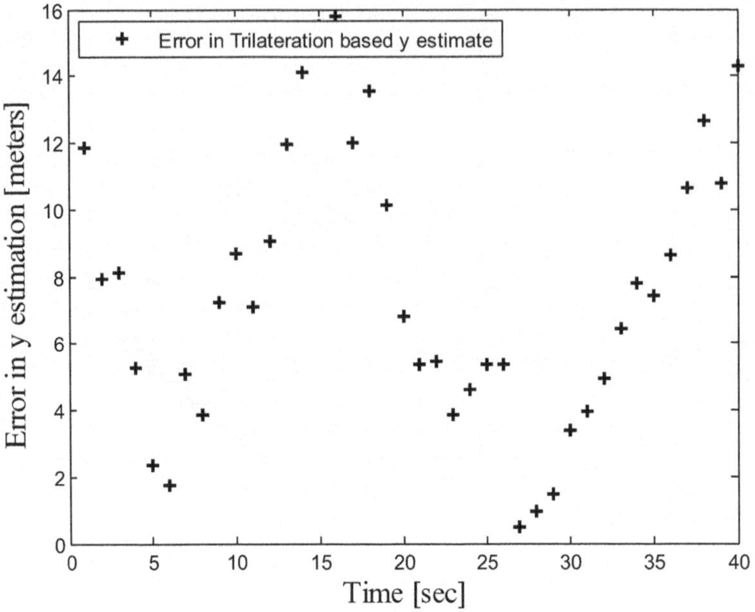

Fig. 4.21 Localization error in *y* estimate (measurement noise = 4 dBm)

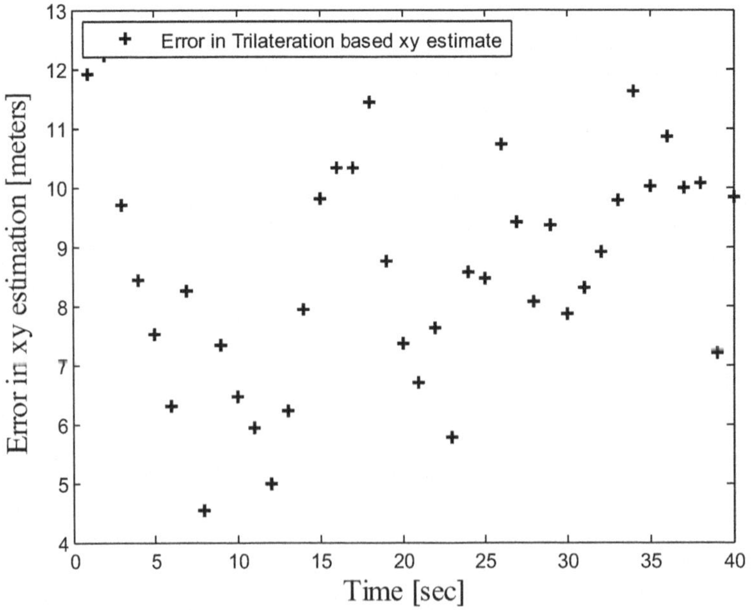

Fig. 4.22 Localization error in *x–y* estimate (measurement noise = 4 dBm)

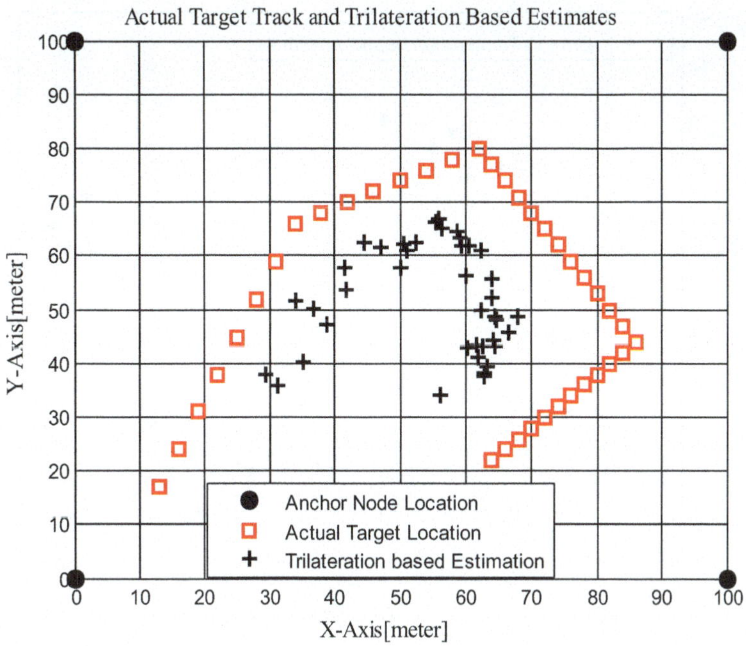

Fig. 4.23 Actual and estimated target trajectories using trilateration algorithm (measurement noise = 5 dBm)

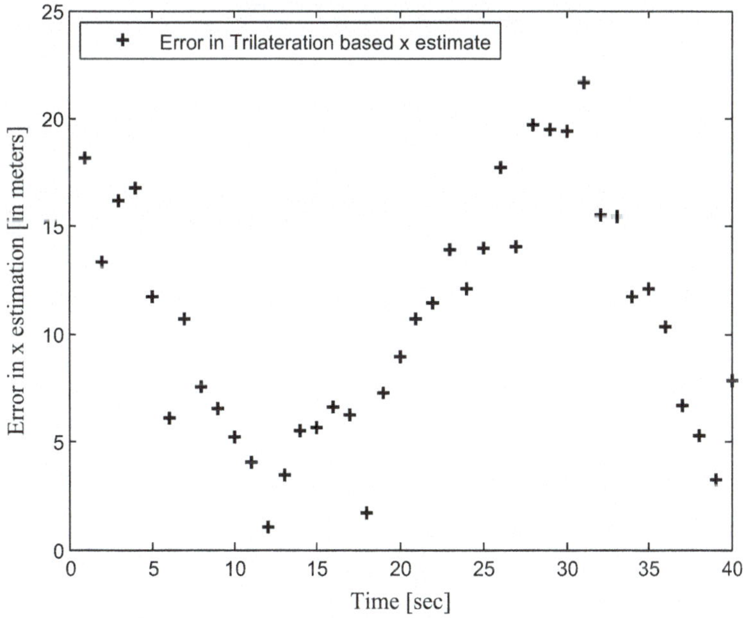

Fig. 4.24 Localization error in *x* estimate (measurement noise = 5 dBm)

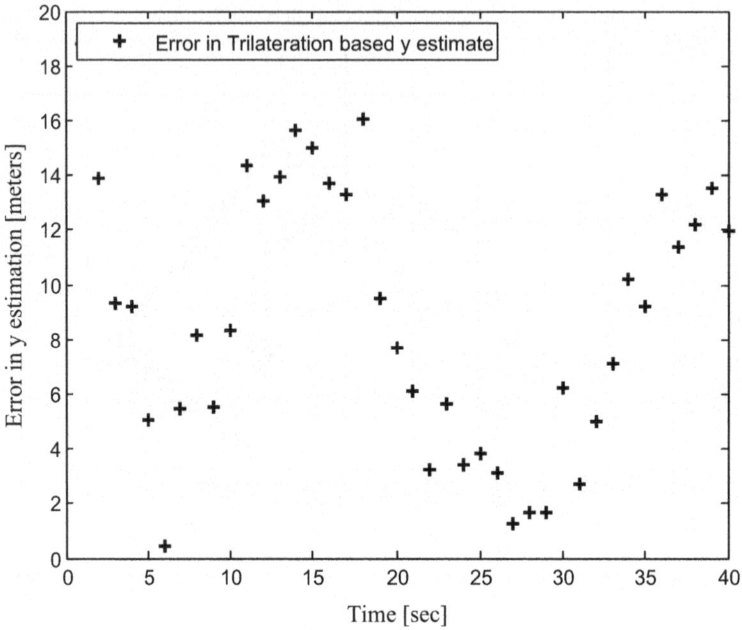

Fig. 4.25 Localization error in *y* estimate (measurement noise = 5 dBm)

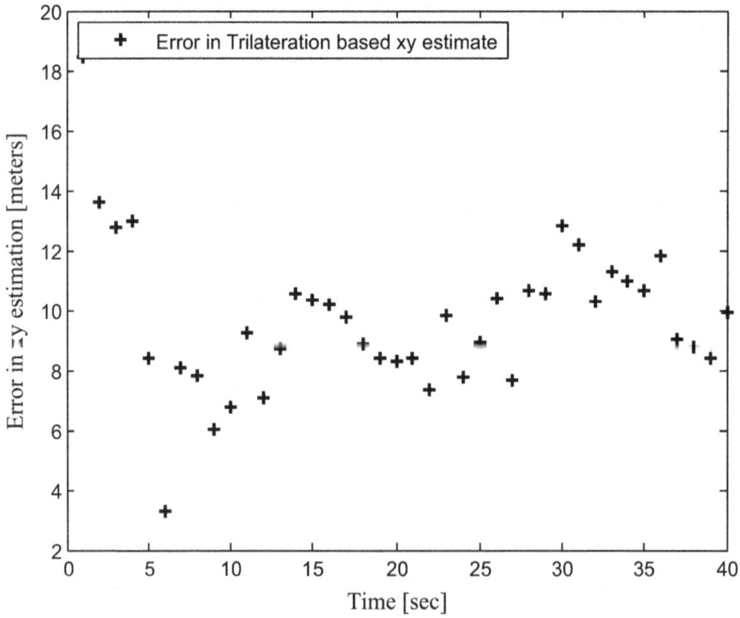

Fig. 4.26 Localization error in *x*–*y* estimate (measurement noise = 5 dBm)

4.4.2 *Case II Results: Testing the Impact of Anchor Density on L&T*

As discussed in the previous section, the objective is to test the impact of anchor density on the trilateration-based L&T algorithm with 4, 6, and 8 anchors (Table 4.4, Figs. 4.27, 4.28, 4.29, 4.30, 4.31, 4.32, 4.33, and 4.34).

The simulation results of Case II reveal that the target L&T performance improves with the rise in the anchor density. The highest tracking performance is observed for anchor density of 8. It is noted that the reduction in RMSE in 8 anchor density case is approximately 15% and 20% as compared to the case of anchor density of 6 and anchor density of 4 (see Table 4.4). Thus, one can claim that the higher the anchor density, the higher the target L&T accuracy will be. Therefore, one can

Table 4.4 Numeric results of Case II: target L&T performance

Evaluation of tracking performance with variation in anchor density (3 dBm noise)			
Sr no	Anchor density	Average localization error (in meters)	RMSE
1	4	6.87	11.02
2	6	6.12	10.50
3	8	5.64	8.92

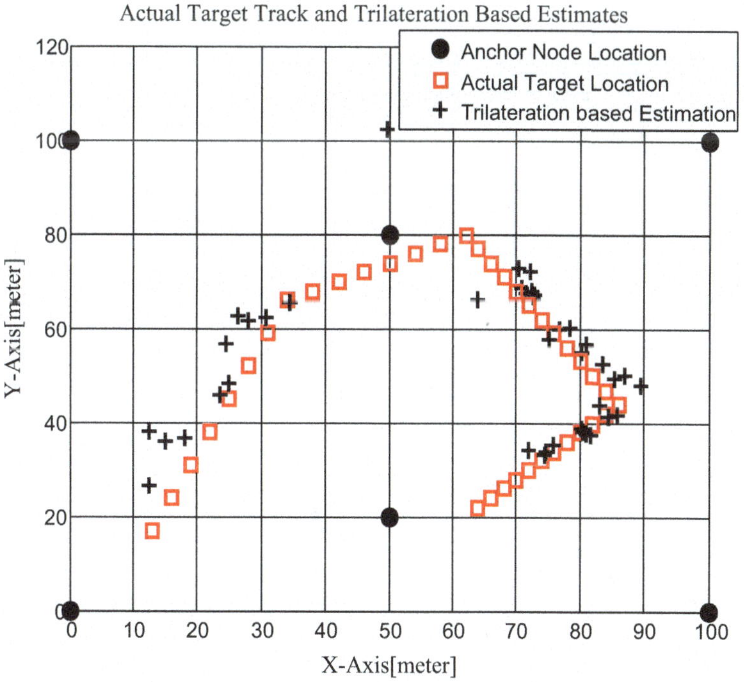

Fig. 4.27 Actual and estimated target trajectories (anchor density = 6)

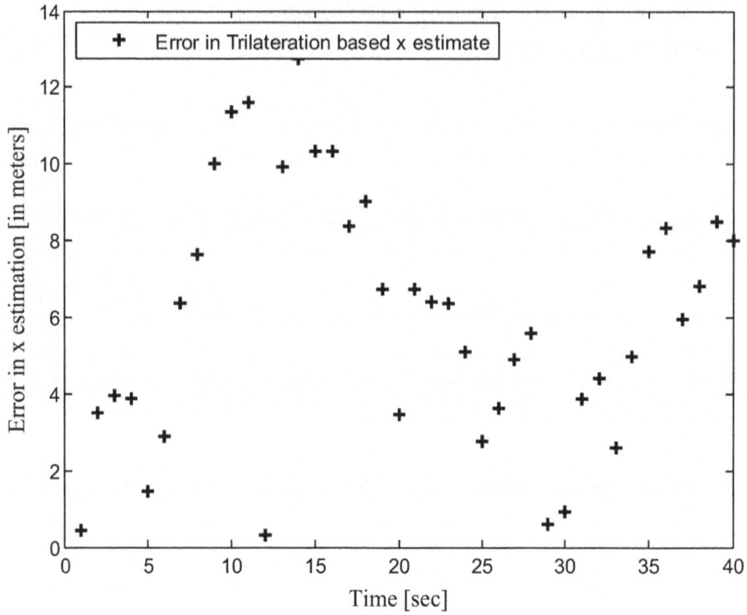

Fig. 4.28 Localization error in x estimate (anchor density = 6)

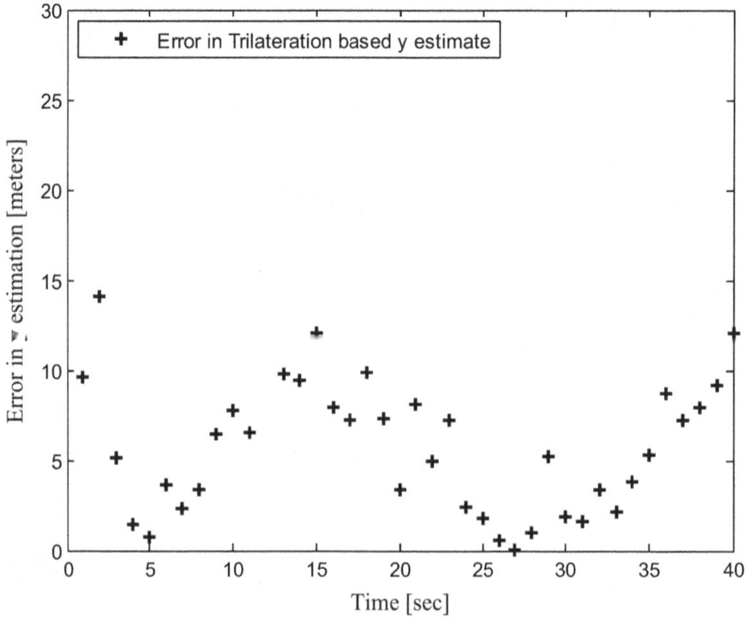

Fig. 4.29 Localization error in y estimate (anchor density = 6)

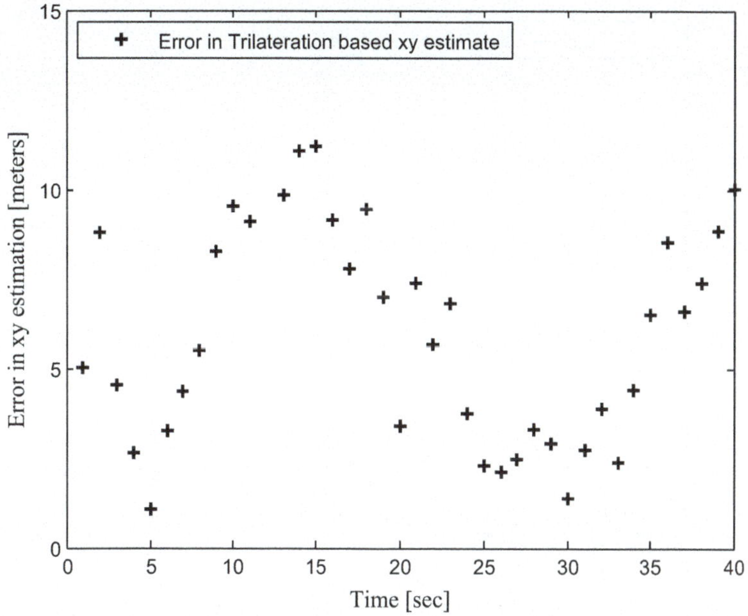

Fig. 4.30 Localization error in x–y estimate (anchor density = 6)

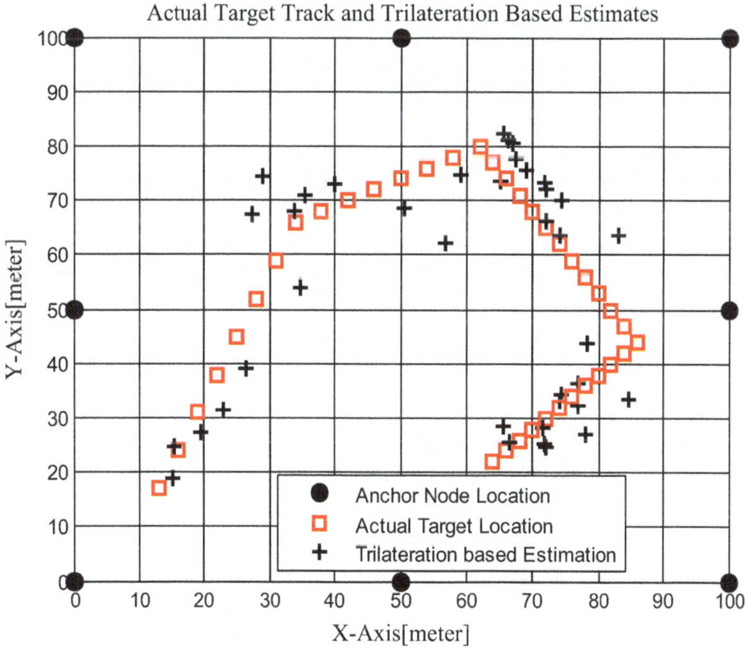

Fig. 4.31 Actual and estimated target trajectories (anchor density = 8)

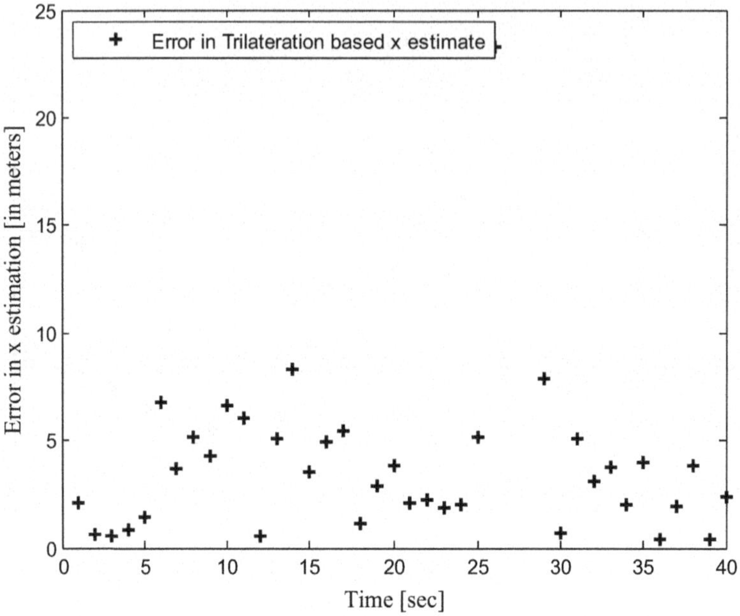

Fig. 4.32 Localization error in *x* estimate (anchor density = 8)

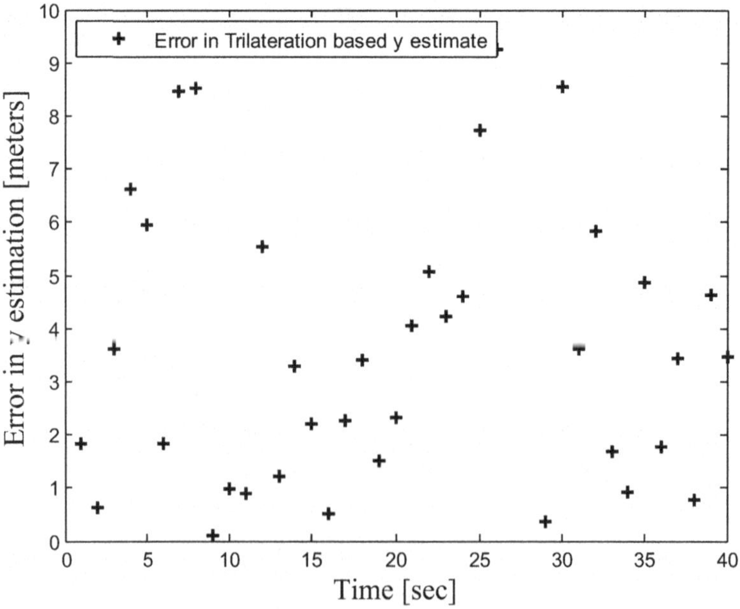

Fig. 4.33 Localization error in *y* estimate (anchor density = 8)

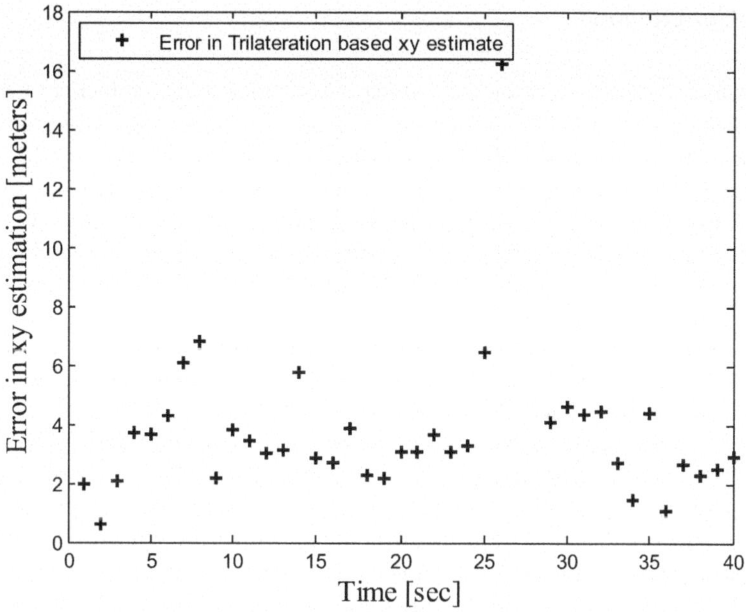

Fig. 4.34 Localization error in x–y estimate (anchor density = 8)

achieve high tracking accuracy by increasing the anchor density but at the cost of increment in hardware and maintenance expenses of the network. It is also quite apparent that the optimum anchor density value for a given test scenario depends on the transmission power (communication range). Increasing the transmission power level may consume more power. It means that there is a trade-off between the tracking performance, transmission power level, and economical budget. The decision of appropriate anchor density and transmission power depends on the dynamicity of the given RF environment and application requirements. This can be a very important research direction in the RSSI-based target L&T.

4.5 Conclusions

This chapter discusses very simple and feasible solution to the problem of L&T of the moving target in WSN with uncertain system dynamics. The target location estimates are obtained with the help of very small anchor density and the RSSI measurements with uncertain measurement noise. The extensive simulation experiments carried out demonstrate high target tracking accuracy, irrespective of the environmental dynamicity and abrupt variations in the velocity of the mobile target. There are many parameters that impact the performance of RSSI-based L&T algorithm, namely, variations in the target velocity, anchor density, and measurement noise in the given RF environment. The higher the anchor density, the higher the target tracking accuracy would be. To study this effect, during simulation experimentation the anchor density is varied from 4 to 8 in steps of 2. To understand the effect of abrupt variations in target velocity, we varied velocity abruptly in the range of −2 to 7 m/s at specific time instances. Whereas to study the impact of environmental dynamicity (variations in fluctuations), the measurement noise in the RSSI is changed from 0 dBm to 5 dBm. The overall target L&T performance is evaluated in terms of the localization error and RMSE. The experimental results prove that the trilateration-based target L&T algorithm fairly estimates the target locations, irrespective of dynamicity in the given RF environment.

The trilateration algorithm has the potential to be used in various applications, such as L&T of mobile robot, person, or assets in indoor and outdoor environmental setting. This research work can be extended in other directions, such as multiple target tracking, discovering the impact of different motion and observation models on localization accuracy, and testing the performance of the tracking algorithm for varying measurement noises and measurement time interval. We also believe that mobile WSN may track the moving target more efficiently than static sensor network. This research may also be extended by assigning mobility to few anchor nodes.

MATLAB Code for Trilateration-Based Target L&T

MATLAB Code for Trilateration Based Target L&T

Main File

```
%% WSN Deployment Setting Parameters
    clear all
    close all
    clc

    networkSize = 100;% we consider a 100by100 area in which target moves

    prompt = 'Enter Number of Anchors : ';
    No_of_Anchors = input(prompt);


if(No_of_Anchors==4)

        anchorLoc   = [0                     0;
                       networkSize           0;
                       0           networkSize;
                       networkSize networkSize];
end

if(No_of_Anchors==6)
    anchorLoc    = [0                    0;
                    networkSize/2        networkSize*0.2;
                    networkSize          0;
                    networkSize          networkSize;
                    networkSize/2        networkSize*0.8;
                    0                    networkSize];
end

if(No_of_Anchors==8)
    anchorLoc    = [0                    0;
                    networkSize/2        0;
                    networkSize          0;
                    networkSize          networkSize/2;
                    networkSize          networkSize;
                    networkSize/2        networkSize;
                    0                    networkSize;
                    0                    networkSize/2];
end


%show anchor Locations
    f1 = figure(1);

plot(anchorLoc(:,1),anchorLoc(:,2),'ko','MarkerSize',8,'lineWidth',2,'MarkerF
aceColor','k');

    grid on
    hold on

% Defining veriables
```

```
    no_of_positions = 40;
    RMSE_rssi= 0;

    error_x_rssi      =      zeros(no_of_positions,1);      error_y_rssi      =
zeros(no_of_positions,1); error_xy_rssi = zeros(no_of_positions,1);

 % Calculate reference RSSI at d0 = 1 meter using Free Space Path Loss Model
 % in Meters

 d0=1;
    Pr0 = RSSI_friss(d0);

    d_test = 20;
    Pr = RSSI_friss(d_test);

 %Calculation of Path Loss Exponent
      n = -(Pr + Pr0)/(10*log(d_test))
      x=10;
      y=10;

 % Generating trajectory for the Target
 for t = 1:no_of_positions
 if(t<9)                    x_v = 3; y_v = 7;
 elseif(t==9 && t<16)       x_v = 4; y_v = 2;
 elseif(t>=16 && t<18)      x_v = 0; y_v = 0;
 elseif(t>=18 && t<30)      x_v = 2; y_v = -3;
 elseif(t>=30 && t<=40)     x_v = -2; y_v = -2;
 end
        x=x+x_v;
        y=y+y_v;

        disp('True Location')
        [x,y]                                          % Actual Target Location
        plot(x,y,'rs','LineWidth',2)

        title('Actual Target Track and Trilateration Based
 Estimates','FontName','Times','FontSize',12);

        ylabel('Y-Axis[meter]','FontName','Times','Fontsize',14);
        xlabel('X-Axis[meter]','FontName','Times','Fontsize',14);

        legend('Anchor Node Location','Actual Target Location','Trilateration
 based Estimation','Location','SouthEast')
        hold on

 % Actual Distances from Anchors required to generate RSSI Values

 if(No_of_Anchors==4)
        d1 = sqrt( x^2 + y^2 );
        d2 = sqrt((100-x)^2 + y^2);
        d3 = sqrt((100-x)^2+ (100-y)^2);
        d4 = sqrt(x^2+ (100-y)^2);
 end

 if(No_of_Anchors==6)
```

```
        d1 = sqrt( x^2 + y^2 );
        d2 = sqrt((networkSize/2 - x)^2 + (networkSize*0.2 - y)^2);
        d3 = sqrt((networkSize - x)^2 + y^2);
        d4 = sqrt((networkSize - x)^2+ (networkSize - y)^2);
        d5 = sqrt((networkSize/2 - x)^2+ (networkSize*0.8 - y)^2);
        d6 = sqrt(x^2+ (networkSize - y)^2);
  end

  if(No_of_Anchors==8)
        d1 = sqrt( x^2 + y^2 );
        d2 = sqrt((networkSize/2 - x)^2 + (y)^2);
        d3 = sqrt((networkSize - x)^2 + y^2);
        d4 = sqrt((networkSize - x)^2+ (networkSize/2 - y)^2);
        d5 = sqrt((networkSize - x)^2+ (networkSize - y)^2);
        d6 = sqrt((networkSize/2 - x)^2+ (networkSize - y)^2);
        d7 = sqrt(x^2+ (networkSize - y)^2);
        d8 = sqrt((x)^2 + (networkSize/2 - y)^2);
  end

  % Generate RSSI Values at 4 Anchor Nodes which are at d1, d2, d3 & d4
distances respectively from Moving Target

  if(No_of_Anchors==4)
        RSS = lognormalshadowing_4(n,d1,d2,d3,d4,Pr0);
        RSS_s = sort(RSS);
  end

  if(No_of_Anchors==6)
        RSS = lognormalshadowing_6(n,d1,d2,d3,d4,d5,d6,Pr0);
        RSS_s = sort(RSS);
  end

  if(No_of_Anchors==8)
        RSS = lognormalshadowing_8(n,d1,d2,d3,d4,d5,d6,d7,d8,Pr0);
        RSS_s = sort(RSS);
  %end
  end

      disp('RSSI Estimated Location')
  if(No_of_Anchors==4)
        mobileLoc_est = trilateration_4(RSS,RSS_s,Pr0,n,networkSize)
        X_T = mobileLoc_est(1);
        Y_T = mobileLoc_est(2);
  end

  if(No_of_Anchors==6)
        mobileLoc_est = trilateration_6(RSS,RSS_s,Pr0,n,networkSize)
        X_T = mobileLoc_est(1);
        Y_T = mobileLoc_est(2);
  end


  if(No_of_Anchors==8)
        mobileLoc_est = trilateration_8(RSS,RSS_s,Pr0,n,networkSize)
        X_T = mobileLoc_est(1);
```

```
        Y_T = mobileLoc_est(2);
end

        plot(X_T,Y_T,'k+','LineWidth',2,'MarkerSize',8);
        hold on

% Error Analysis of algorithm
% ---> Part 1 : for Trilateration based Technique
        RMSE_rssi = RMSE_rssi + ((X_T - x)^2 + (Y_T - y)^2);

% ---> Part 2 : Calculation of Absolute Errors

        error_x_rssi(t) = abs((x - X_T));
        error_y_rssi(t) = abs((y - Y_T));
        error_xy_rssi(t) =((error_x_rssi(t) + error_y_rssi(t))/2);
end
% Average Error in x & y  coordinates
       avg_error_xy_rssi = 0;

for t = 1:no_of_positions
 avg_error_xy_rssi=avg_error_xy_rssi+(error_xy_rssi(t)/no_of_positions);
end

      avg_error_xy_rssi

% Average Error in x & y  coordinates
       RMSE_rssi = sqrt(RMSE_rssi/no_of_positions)


% Plotting Absolute Errors of KF & UKF based Tracking

f2 = figure(2);
for t =1:no_of_positions

    plot(t,error_x_rssi(t),'k+','Linewidth',2)
    xlabel('Time [sec]','FontName','Times','Fontsize',14)
    ylabel('Error in x estimation [in
meters]','FontName','Times','Fontsize',14)
    hold on

end
  legend('Error in Trilateration based x estimate','Location','NorthWest')

  f3 = figure(3);
for t =1:no_of_positions

    plot(t,error_y_rssi(t),'k+','Linewidth',2)
    xlabel('Time [sec]','FontName','Times','Fontsize',14)
    ylabel('Error in y estimation
[meters]','FontName','Times','Fontsize',14)

  hold on
end
  legend('Error in Trilateration based y estimate','Location','NorthWest')
```

```
  f4 = figure(4);
for t =1:no_of_positions

    plot(t,error_xy_rssi(t),'k+','Linewidth',2)
    xlabel('Time [sec]','FontName','Times','Fontsize',14)
    ylabel('Error in xy estimation
[meters]','FontName','Times','Fontsize',14)

  hold on
end
  legend('Error in Trilateration based xy estimate','Location','NorthWest')

  linearity_test(n,Pr0);

  f7 = figure(7);
for t =1:no_of_positions

if(t<9)                 x_v = 3; y_v = 7;
elseif(t==9 && t<16)    x_v = 4; y_v = 2;
elseif(t>=16 && t<18)   x_v = 0; y_v = 0;
elseif(t>=18 && t<30)   x_v = 2; y_v = -3;
elseif(t>=30 && t<=40)  x_v = -2; y_v = -2;
end

    plot(t,x_v,'b+','Linewidth',2)
    xlabel('Time, [sec]','FontName','Times','Fontsize',14)
    ylabel( 'Target Velocity in x Direction
[meters/sec]','FontName','Times','Fontsize',14)
    title('Target Velocity in x direction wrt
Time','FontName','Times','FontSize',12);
    hold on

end

  f8 = figure(8);
for t =1:no_of_positions

if(t<9)                 x_v = 3; y_v = 7;
elseif(t==9 && t<16)    x_v = 4; y_v = 2;
elseif(t>=16 && t<18)   x_v = 0; y_v = 0;
elseif(t>=18 && t<30)   x_v = 2; y_v = -3;
elseif(t>=30 && t<=40)  x_v = -2; y_v = -2;
end

    plot(t,y_v,'b+','Linewidth',2)
    xlabel('Time [sec]','FontName','Times','Fontsize',14)
    ylabel('Target Velocity in y Direction
[meters/sec]','FontName','Times','Fontsize',14)
    title('Target Velocity in y direction wrt
Time','FontName','Times','FontSize',12);
    hold on
end
```

lognormalshadowing_4.m

```
function [ RSS ] = lognormalshadowing_4( n,d1,d2,d3,d4,Pr0 )
%UNTITLED5 Summary of this function goes here
%   Detailed explanation goes here
%clear all
        std_meas_noise = 3;  % standard deviation 0f measurement noise in db


        RSS_d1= -(10*n* log10(d1) + Pr0) + (std_meas_noise + 1.*randn);
        RSS_d2= -(10*n* log10(d2) + Pr0) + (std_meas_noise + 1.*randn);
        RSS_d3= -(10*n* log10(d3) + Pr0) + (std_meas_noise + 1.*randn);
        RSS_d4= -(10*n* log10(d4) + Pr0) + (std_meas_noise + 1.*randn);

        RSS = [RSS_d1, RSS_d2, RSS_d3, RSS_d4];
end
```

trilateration_4.m

```
function [ mobileLoc_est ] = trilateration_4( RSS,RSS_s,Pr0,n,networkSize )
%UNTITLED6 Summary of this function goes here
%   Detailed explanation goes here
% Select   highest  three  RSSI  Values   &  Calculate  distances  using  d  =
antilog(Pr0-RSSI)/(10*n)

            d1_est = 10^(-(Pr0+RSS_s(4))/(10*n));
            d2_est = 10^(-(Pr0+RSS_s(3))/(10*n));
            d3_est = 10^(-(Pr0+RSS_s(2))/(10*n));

if (RSS_s(4) == RSS(1))
                X1 = 0; Y1 = 0;
elseif (RSS_s(4) == RSS(2))
                X1 = networkSize; Y1 = 0;
elseif (RSS_s(4) == RSS(3))
                X1 = networkSize; Y1 = networkSize;
elseif (RSS_s(4) == RSS(4))
                X1 = 0; Y1 = networkSize;
end

if (RSS_s(3) == RSS(1))
                X2 = 0; Y2 = 0;
elseif (RSS_s(3) == RSS(2))
                X2 = networkSize; Y2 = 0;
elseif (RSS_s(3) == RSS(3))
                X2 = networkSize; Y2 = networkSize;
elseif (RSS_s(3) == RSS(4))
                X2 = 0; Y2 = networkSize;
end

if (RSS_s(2) == RSS(1))
                X3 = 0; Y3 = 0;
elseif (RSS_s(2) == RSS(2))
                X3 = networkSize; Y3 = 0;
elseif (RSS_s(2) == RSS(3))
                X3 = networkSize; Y3 = networkSize;
elseif (RSS_s(2) == RSS(4))
                X3 = 0; Y3 = networkSize;
end
```

```
% Trilateration Algorithm (x - xi)^2 + (y - yi)^2 = di^2
% d1^2 = (x1-x)^2 + (y1-y)^2   similarly write for d2 & d3
% Calculate A , B, C and then Target coordinates are given as follows

        A = X1^2 + Y1^2 - d1_est^2;
        B = X2^2 + Y2^2 - d2_est^2;
        C = X3^2 + Y3^2 - d3_est^2;

        X32 = (X3 - X2);
        Y32 = (Y3 - Y2);
        X21 = (X2 - X1);
        Y21 = (Y2 - Y1);
        X13 = (X1 - X3);
        Y13 = (Y1 - Y3);

        X_T = (A*Y32 + B*Y13 + C*Y21) / (2*( X1*Y32 + X2*Y13 + X3*Y21));
% X coordinate of target

        Y_T = (A*X32 + B*X13 + C*X21) / (2*( Y1*X32 + Y2*X13 + Y3*X21));
% Y coordinate of target

        mobileLoc_est = [ X_T, Y_T];

end
```

linearity_test.m

```
function [  ] = linearity_test( n,Pr0 )
%UNTITLED Summary of this function goes here
%   Detailed explanation goes here

        std_meas_noise = 1;  % standard deviation 0f measurement noise in db
for d=0:0.8:100
        RSS = -(10*n* log10(d) + Pr0) + std_meas_noise*randn;
        f5 = figure(5);
        plot(d,RSS,'r+','LineWidth',2)
        xlabel('Distance, [m]','FontName','Times','Fontsize',14)
        ylabel('RSSI Measurement [dbm]','FontName','Times','Fontsize',14)
        hold on
end
        disp('Figure 6 shows non linearity in RSSI values with distance')
        text(8,10,'\leftarrow RSSI Curve')
        title('RSSI versus Distance Curve','FontName','Times','FontSize',14)

        f6 = figure(6);
        d=10;
for t=1:10
         RSS(t) = -(10*n* log10(d) + Pr0) + std_meas_noise*randn;
         plot(t,RSS(t),'r+','LineWidth',2)
         xlabel('Time, [sec]','FontName','Times','Fontsize',14)
         ylabel('RSSI value, [dbm]','FontName','Times','Fontsize',14)
         hold on
 end

        title('Fluctuations in RSSI Measurements at Distance of 10 meter wrt
Time','FontName','Times','FontSize',12)
 end
```

References

1. E.E.L. Lau, W.Y. Chung, Enhanced RSSI-based real-time user location tracking system for indoor and outdoor environments, in *International Conference on Convergence Information Technology* (2007). https://doi.org/10.1109/ICCIT.2007.4420422
2. A.S. Paul, E.A. Wan, RSSI-based indoor localization and tracking using sigma-point kalman smoothers. IEEE J. Sel. Top. Signal Process. **3**, 860–873 (2009). https://doi.org/10.1109/JSTSP.2009.2032309
3. S.R. Jondhale, R.S. Deshpande, Kalman filtering framework-based real time target tracking in wireless sensor networks using generalized regression neural networks. IEEE Sensors J. **19**(1), 224–233 (2019). https://doi.org/10.1109/JSEN.2018.2873357
4. S. Vougioukas, H.T. Anastassiu, C. Regen, M. Zude, Influence of foliage on radio path losses (PLs) for Wireless Sensor Network (WSN) planning in orchards. Biosyst. Eng. **114**, 454–465 (2013). https://doi.org/10.1016/j.biosystemseng.2012.08.011
5. T.K. Sarkar, Z. Ji, K. Kim, A. Medouri, M. Salazar-Palma, A survey of various propagation models for mobile communication. IEEE Antennas Propag Mag **45**(3), 51–82 (2003). https://doi.org/10.1109/MAP.2003.1232163
6. H. Wu, L. Zhang, Y. Miao, The propagation characteristics of radio frequency signals for wireless sensor networks in large-scale farmland. Wirel. Pers. Commun. **95**(4), 3653–3670 (2017). https://doi.org/10.1007/s11277-017-4018-5
7. D. Balachander, T.R. Rao, G. Mahesh, RF propagation investigations in agricultural fields and gardens for wireless sensor communications, in *IEEE Conference on Information Communication Technologies* (2013). https://doi.org/10.1109/CICT.2013.6558195
8. S.R. Jondhale, R.S. Deshpande, Kalman filtering framework based real time target tracking in wireless sensor networks using generalized regression neural networks. IEEE Sensors J (2018). https://doi.org/10.1109/JSEN.2018.2873357
9. S.R. Jondhale, R.S. Deshpande, GRNN and KF framework based real time target tracking using PSOC BLE and smartphone. Ad Hoc Netw (2019). https://doi.org/10.1016/j.adhoc.2018.09.017
10. S.R. Jondhale, R.S. Deshpande, Modified Kalman filtering framework based real time target tracking against environmental dynamicity in wireless sensor networks. Ad Hoc Sensor Wirel. Netw **40**, 119–143 (2018)

Chapter 5
KF-Based Target L&T Using RSSI

5.1 System Assumptions and Design of KF-Based L&T

The proposed system contains a set of stationary anchor nodes at some fixed locations. These anchor nodes are supposed to be placed in 100 m by 100 m area, whereas a single target moves through this area, as shown in Figs. 5.3, 5.5, 5.7, and 5.9. The base station is assumed to be located outside the WSN area. The proposed L&T system is supposed to run for the total period of T, which are divided into many time slots, each of which is denoted by notation dt. The mobile target is supposed to have a WSN node, and it broadcasts RF signal to all the anchor nodes on a time instance k. That means, it is assumed that the anchor nodes act as the receivers and the target is a transmitter in the proposed experimental setup. It can be considered as a case of cooperative L&T. Each anchor node computes their distances from the moving target using the received RSSI measurements. The detailed mathematics behind these computations is explained with the help of the equations given in this section. The anchor nodes send these distances and their coordinates to the base station, kept outside the WSN area considered for the proposed system. The base station then selects the lowest three distances out of them. The base station is assumed to be a laptop (Processor: Core i3, 1.89 GHz, 2 GB RAM). Upon receiving the details from the anchor nodes, the base station runs the trilateration algorithm to estimate the moving target location for each of the sampling interval. The sensor node communication range is considered 100 m during simulation experiments. The transmission power is assumed to be 1 mW, whereas the receiver and transmitter antenna gains are 1 dBm.

In this chapter, the simulation experiments are divided into three cases as discussed below:

S. R. Jondhale et al., *Received Signal Strength Based Target Localization and Tracking Using Wireless Sensor Networks*, EAI/Springer Innovations in Communication and Computing, https://doi.org/10.1007/978-3-030-74061-0_5

- *Case I: Testing the proposed trilateration+KF and trilateration+UKF L&T algorithms for nonlinear target trajectory with anchor density = 6 and WSN area of 100 m × 100 m.*
- *Case II: Testing the proposed trilateration+KF and trilateration+UKF L&T algorithms for linear target trajectory with anchor density = 8 and WSN area of 100 m × 100 m.*
- *Case III: Testing the proposed trilateration+KF and trilateration+UKF L&T algorithms for linear target trajectory with anchor density = 8 and WSN area of 200 m × 200 m.*

In Case I the target is assumed to go through the abrupt variations in the velocity during T sec as per Eqs. (5.1)–(5.5). These variations in velocity are depicted in Figs. 5.1 and 5.2. The negative value for the target velocity means that the target motion is toward a location of small coordinate value as compared to the location for previous instance of time.

$$\dot{x}_k = 3, \quad \dot{y}_k = 7, \qquad \text{for } 0 < k < 9 \text{ s}, \tag{5.1}$$

$$\dot{x}_k = 4, \quad \dot{y}_k = 2, \qquad \text{for } 9 \le k \le 16 \text{ s}, \tag{5.2}$$

$$\dot{x}_k = 0, \quad \dot{y}_k = 0, \qquad \text{for } 16 \le k \le 18 \text{ s}, \tag{5.3}$$

$$\dot{x}_k = 2, \quad \dot{y}_k = -3, \qquad \text{for } 18 \le k \le 30 \text{ s}. \tag{5.4}$$

$$\dot{x}_k = -2, \quad \dot{y}_k = -2, \qquad \text{for } 30 \le k \le 40 \text{ s}. \tag{5.5}$$

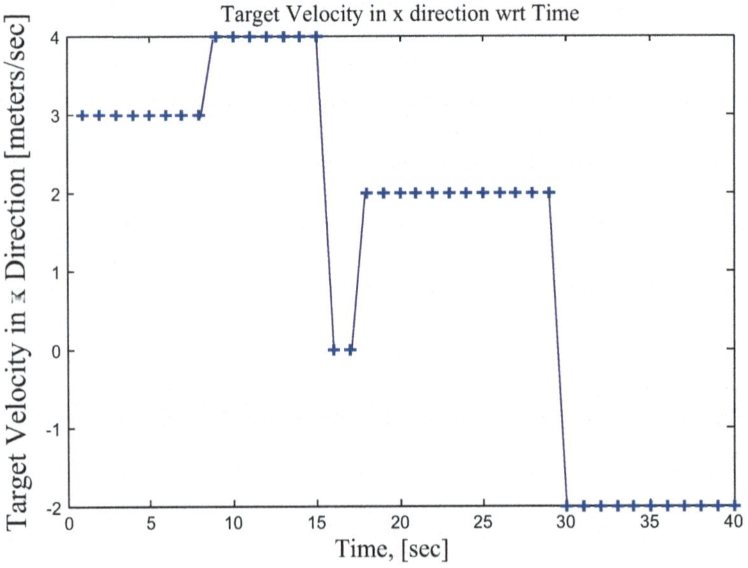

Fig. 5.1 Abrupt variations in target velocity along x direction (Case I)

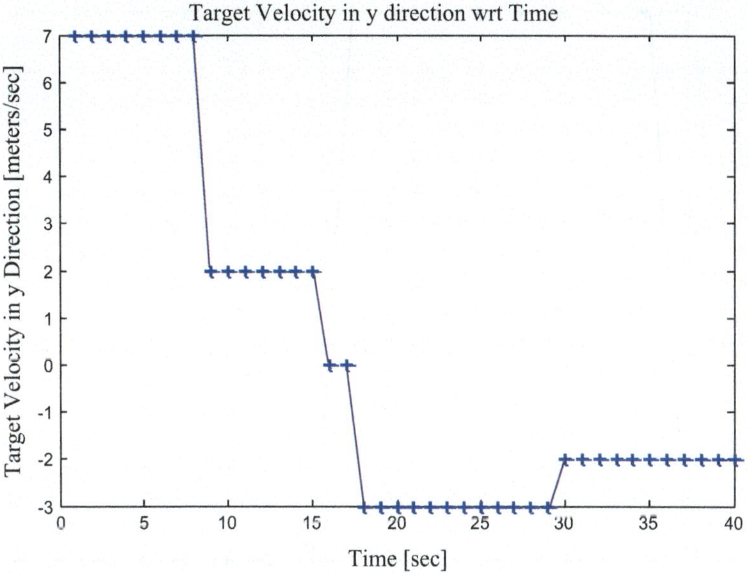

Fig. 5.2 Abrupt variations in target velocity along *y* direction (Case I)

In Case II the target is assumed to go through the abrupt variations in the velocity during *T* sec as per Eqs. (5.6)–(5.9). This variation in velocity is illustrated in Figs. 5.3 and 5.4.

$$\dot{x}_k = 5, \quad \dot{y}_k = 0, \qquad \text{for } 0 < k < 1 \text{ s}, \tag{5.6}$$

$$\dot{x}_k = 10, \quad \dot{y}_k = 0, \qquad \text{for } 1 \le k \le 8 \text{ s}, \tag{5.7}$$

$$\dot{x}_k = 5, \quad \dot{y}_k = 5, \qquad \text{for } 8 \le k \le 9 \text{ s}, \tag{5.8}$$

$$\dot{x}_k = 0, \quad \dot{y}_k = 10, \qquad \text{for } 9 \le k \le 16 \text{ s}. \tag{5.9}$$

In Case III the target is assumed to go through the abrupt variations in the velocity during *T* sec as per Eqs. (5.10)–(5.12). This variation in velocity is illustrated in Figs. 5.5 and 5.6.

$$\dot{x}_k = 6, \quad \dot{y}_k = 0, \qquad \text{for } 0 < k < 1 \text{ s}, \tag{5.10}$$

$$\dot{x}_k = 12, \quad \dot{y}_k = 0, \qquad \text{for } 1 \le k \le 17 \text{ s}, \tag{5.11}$$

$$\dot{x}_k = 0, \quad \dot{y}_k = 12, \qquad \text{for } 17 \le k \le 32 \text{ s}, \tag{5.12}$$

This chapter follows LNSM for simulation experiments. The RSSI ($z_{lj,k}$) at the node N_l with coordinates (x_{lk}, y_{lk}) at time instance *k*, when transmitted from the node N_j with coordinates (x_{jk}, y_{jk}), is based on Eq. (5.13) as given below [1–3]:

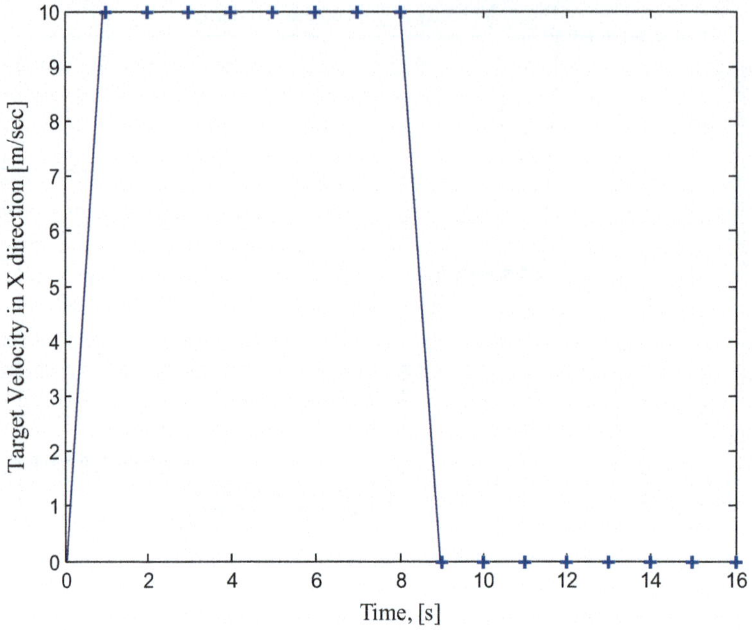

Fig. 5.3 Abrupt variation in target velocity in x direction during motion (Case II)

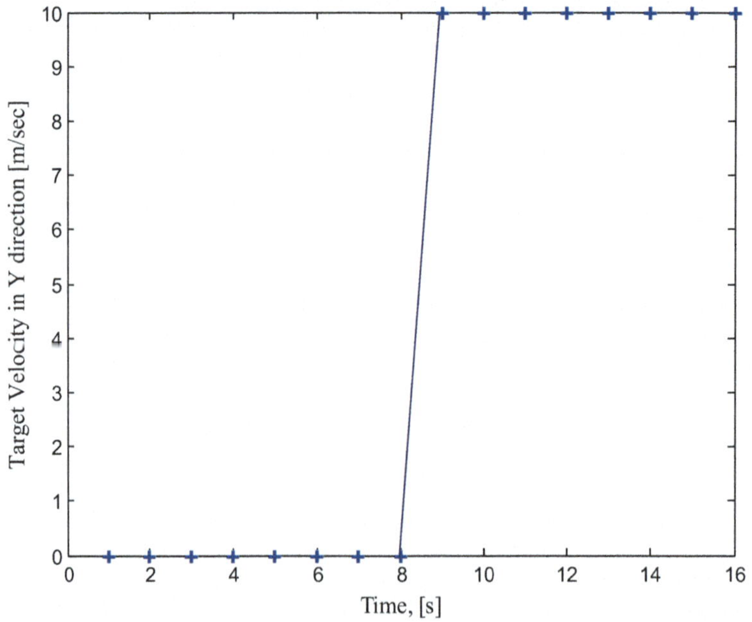

Fig. 5.4 Abrupt variation in target velocity in y direction during motion (Case II)

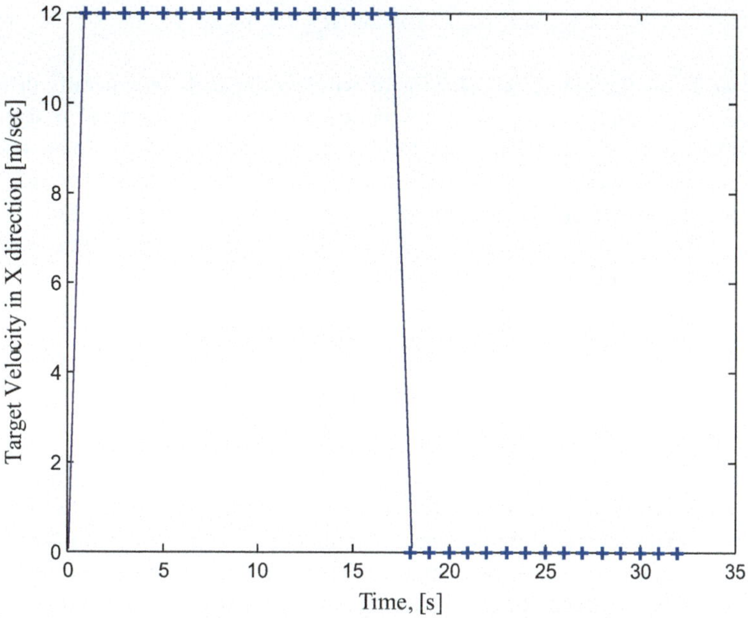

Fig. 5.5 Abrupt variation in target velocity in x direction during motion (Case III)

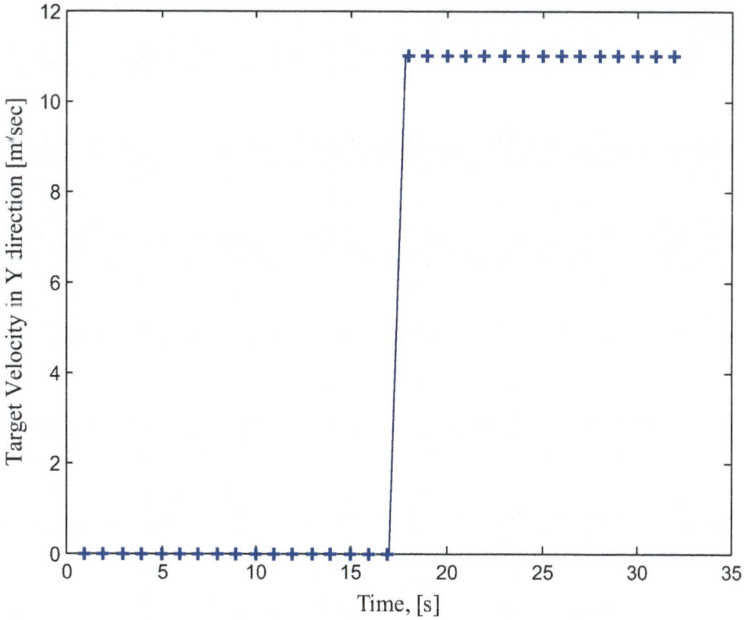

Fig. 5.6 Abrupt variation in target velocity in y direction during motion (Case III)

$$z_{lj,k} = P_r\left(d_0\right) - 10n\log\left(d_{lj,k} / d_0\right) + X_\sigma \ , \tag{5.13}$$

As we know the nature or behavior of wireless channels is generally different because of variations in the obstacles between the receiver and the transmitter; hence, the values of n and $P_r(d_0)$ need to be chosen carefully [4–6]. To address the environmental dynamicity in the RF environment, average values of n (n_{avg}) and $P_r(d_0)$ are computed in advance during system calibration step (see Eqs. (5.14)–(5.19)). Given three distances (d_1, d_2 and d_3), and three RSSI measurements (z_3, z_2 and z_3), the value of n (n_{avg}) is calculated as given below:

$$z_1 = P_r\left(d_0\right) - 10n_1\log\left(d_1 / d_0\right) + X_\sigma \ , \tag{5.14}$$

$$z_2 = P_r\left(d_0\right) - 10n_2\log\left(d_2 / d_0\right) + X_\sigma \ , \tag{5.15}$$

$$z_3 = P_r\left(d_0\right) - 10n_3\log\left(d_3 / d_0\right) + X_\sigma \ . \tag{5.16}$$

where (n_1, n_2 and n_3) are path loss exponents, corresponding to distances d_1, d_2 and d_3, respectively. By rearranging and subtracting the above equations, the values of n_1, n_2 and n_3 can easily be computed. Thereafter, the average path loss exponent n_{avg} can be calculated by taking average of n_1, n_2 and n_3 as given below:

$$n_{avg} = \left(n_1 + n_2 + n_3\right) / 3 \tag{5.17}$$

Therefore, Eq. (5.13) is modified as:

$$z_{lj,k} = P_r\left(d_0\right) - 10n_{avg}\log\left(d_{lj,k} / d_0\right) + X_\sigma . \tag{5.18}$$

The value of $P_r(d_0)$ can then be calculated using Eq. (5.18) by utilizing the value of RSSI $z_{lj,k}$ for a given distance $d_{lj,k}$ and value of n_{avg}.

$$P_r\left(d_0\right) = z_{lj,k} + 10n_{\text{avg}}\log\left(d_{lj,k} / d_0\right) - X_\sigma \ . \tag{5.19}$$

Also the distance between the transmitter and the receiver can be computed as given below:

$$d_{lj,k} = d_0 10^{\left(P_r\left(d_0\right) - z_{lj,k} + X_\sigma\right) / 10n_{\text{avg}}} \ . \tag{5.20}$$

The research works in all the above cases follow the CV mobility model of the target. The state of the moving target at time instance k is defined by the target state vector $X_k = \left(x_k, y_k, \dot{x}_k, \dot{y}_k\right)$, wherein variables x_k and y_k specify the locations and $\dot{x}_k$ and $\dot{y}_k$ specify the velocities along x and y directions, respectively. The target motion for the research work in this chapter is defined below:

$$x_k = x_{k-1} + \dot{x} \ dt \ , \tag{5.21}$$

$$y_k = y_{k-1} + \dot{y} \ dt \ . \tag{5.22}$$

Table 5.1 Case I simulation parameters

Notation	Parameter	Magnitude
X_0	Target state vector at $k = 0$	[12 18 0 0]
–	Anchor density	6
dt	Discretization time step	1 s
T	Simulation period	40 s
X_σ	Normal random variable	~N (3, 1)
f	Frequency	2.4 GHz
n_{avg}	Average path loss exponent	3.40

Table 5.2 Case II and Case III simulation parameters

Symbol	Parameter	Magnitude
X_0	Initial target state at $k = 0$	[10 20 0 0]
–	Anchor density	8
dt	Discretization time step	1 s
T	Simulation period	16 s (Case II) & 32 s (Case III)
X_σ	Normal random variable	~N (3, 1)
f	Frequency	2.4 GHz
n_{avg}	Average path loss exponent	3.40

where dt is discrete time step between two consecutive time instance during target tracking. The dt is also called as discretization time step. The initial values of R, $P\,and\,Q$ matrices considered for this implementation are [1–3]:

$$R = \begin{bmatrix} 2.2 & 0 & 0 & 0 \\ 0 & 1.2 & 0 & 0 \\ 0 & 0 & 0.9 & 0 \\ 0 & 0 & 0 & 0.5 \end{bmatrix}, \; P = \begin{bmatrix} 0.25 & 0 & 0 & 0 \\ 0 & 0.4 & 0 & 0 \\ 0 & 0 & 0.2 & 0 \\ 0 & 0 & 0 & 0.01 \end{bmatrix}, \; Q - I_{4\times4} \, . \tag{5.23}$$

The initial target state vectors for Case II and Case III are assumed to be the same; however, that for Case I is different from Case II and Case III. The detailed simulation parameters for these cases are as given below in Tables 5.1 and 5.2:

5.2 Flow of Trilateration+KF and Trilateration+UKF-Based L&T Algorithms

The simulation during one time instance k contains three steps. The first step is the environmental calibration (off-line step) (computation of n_{avg} and $P_r(d_0)$ empirically) [1–3]. The second part is the computation of distances, whereas the third part is the application of an input (three smallest distances between the anchors and the mobile target (obtained in the second part) and their coordinates) to the trilateration

Table 5.3 Trilateration+KF and trilateration+UKF-based target L&T algorithms

I. Off-line environmental calibration
Step 1: Determine n_{avg} and $P_r(d_0)$
II. For $k = 0$
Step 2: Anchor nodes receive the RSSIs for every kth instance of time from the target, which are to be utilized to compute their distances ($d_1, d_2, \ldots d_n$) from the moving target
Step 3: Anchors share the calculated distances ($d_1, d_2, \ldots d_n$) and corresponding coordinates to the base station. The base station selects the smallest three distances and corresponding coordinates of anchors
III. Location estimation at the base station
Step 4: The base station executes the trilateration to estimate x and y locations of the target using step 3 inputs
Step 5:The trilateration estimated location is refined using the trilateration+KF and trilateration+UKF algorithms. That means the results of step 4 is used as an input to KF and UKF algorithms in step 5
For $k = 1, 2, 3, 4 \ldots T$
Step 6: Repeat step 1 to step 5 for each next time instance k
Step 7: Calculate the RMSE and average localization error for the trilateration+KF algorithm and trilateration+UKF algorithm at the base station

algorithm. The obtained simulation results are based on the average of more than 50 trials of the proposed trilateration algorithm. The detailed flow of the trilateration-based L&T algorithm during time instance k is explained in detail in Table 5.3.

5.3 Performance Metrics for Assessment of L&T Performance

To assess the localization performance of the trilateration+KF and trilateration+UKF algorithms, two performance metrics are used, namely, RMSE and average localization error. These metrics represent the average estimation error in the target location $(\hat{x}_k, \hat{y}_k)$ and how much close the estimated target location $(\hat{x}_k, \hat{y}_k)$ and actual target location (x_k, y_k), respectively. These two metrics collectively represent a measure of target L&T accuracy [7–10]. The lower the values of these parameters, the higher the target L&T accuracy would be. For each of the time instance k, the error in x estimate $(\hat{x}_k - x_k)$ and error in y estimate $(\hat{y}_k - y_k)$ for the trilateration+KF and trilateration+UKF algorithms are computed for Case I, Case II, and Case III. For each simulation run, the RMSE and average localization error are computed with the help of Eqs. (5.24) and (5.25), respectively [1–3].

1. *Average Localization Error*

$$\text{Average Localization Error} = \frac{1}{T}\sum_{k=1}^{T}\frac{(\hat{x}_k - x_k) + (\hat{y}_k - y_k)}{2}, \tag{5.24}$$

2. *RMSE*

$$\mathrm{RMSE} = \sqrt{\frac{1}{T}\sum_{k=1}^{T}\frac{(\hat{x}_k - x_k)^2 + (\hat{y}_k - y_k)^2}{2}}. \tag{5.25}$$

5.4 Discussion on Results

5.4.1 Case I Results

Testing the proposed trilateration+KF and trilateration+UKF L&T algorithms for nonlinear target trajectory with anchor density = 6 and WSN area of 100 m × 100 m.

In Case I, the anchor nodes are kept at locations in WSN-defined area as given in Table 5.4:

In Case I, the target starts moving from location (12, 18) and stops at (63, 22) as shown in Fig. 5.7. These figures depict the actual and the estimated target tracks using the trilateration, trilateration+KF, and trilateration+UKF algorithms. The color combinations for various terms in Fig. 5.7 are as given below:

For anchor nodes: black filled squares.
For actual target location at any time instance k: red unfilled squares.
For estimated location using trilateration: red plus symbol.
For estimated location using trilateration+KF: blue plus symbol.
For estimated location using trilateration+UKF: black circle symbol.

The numeric results in Case I are presented in Table 5.5. From these results it is clear that the RMSE and average localization error are highest for trilateration-based estimation, moderate for trilateration+KF-based estimation, and lowest for trilateration+UKF-based estimation. That means the trilateration+UKF-based target L&T algorithm is superior in L&T performance than the rest of the two L&T algorithms. The % reductions in the RMSE in the trilateration+UKF-based L&T algorithm as compared to the trilateration+KF algorithm and trilateration algorithm are 68% and 91%, respectively. Thus, we can easily claim that KF and UKF algorithms are able to improve the target L&T performance of the trilateration-based estimation technique further. In other words, the proposed trilateration+KF and

Table 5.4 Anchor node deployment in Case I

Anchor node	2-D location (in meters)
1	(0, 0)
2	(100, 0)
3	(0, 100)
4	(100, 100)
5	(50, 20)
6	(50, 80)

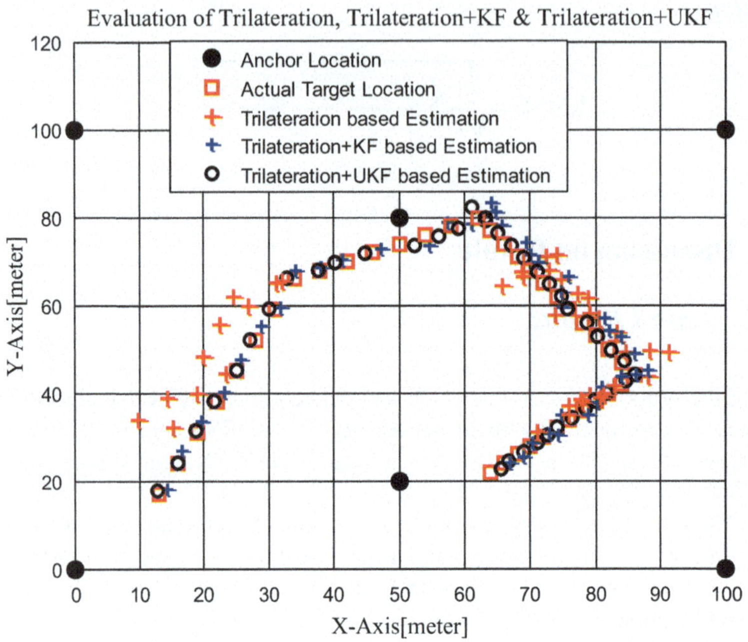

Fig. 5.7 Actual and estimated target tracks using trilateration, trilateration+KF, and trilateration+UKF algorithms (Case I)

Table 5.5 Numeric results obtained using trilateration, trilateration+KF, and trilateration+UKF target L&T algorithms (Case I)

Case I, 3 dBm			
Sr no	Measurement noise (dBm)	Avg localization error (in meters)	RMSE
1	Trilateration	6.6669	11.3263
2	Trilateration+KF	1.7478	3.0062
3	Trilateration+UKF	0.5862	0.9459

trilateration+UKF algorithms can better deal with the abrupt changes in the target velocity as well as the environmental dynamicity (Figs. 5.8, 5.9, and 5.10).

5.4.2 *Case II Results*

Testing the proposed trilateration+KF and trilateration+UKF L&T algorithms for linear target trajectory with anchor density = 8 and WSN area of 100 m × 100 m.

In Case II, the anchor nodes are kept at locations in WSN-defined area as given in Table 5.6.

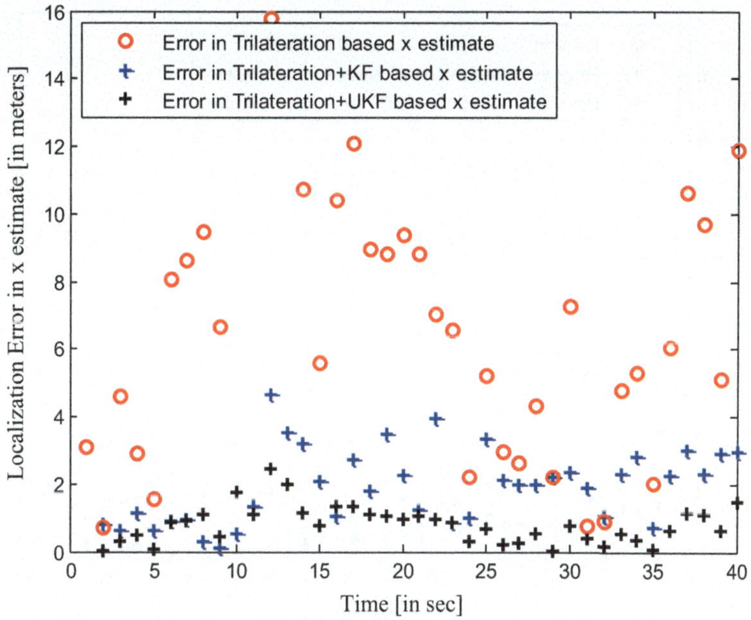

Fig. 5.8 Target localization error in x estimate with trilateration, trilateration+KF, and trilateration+UKF algorithms (Case I)

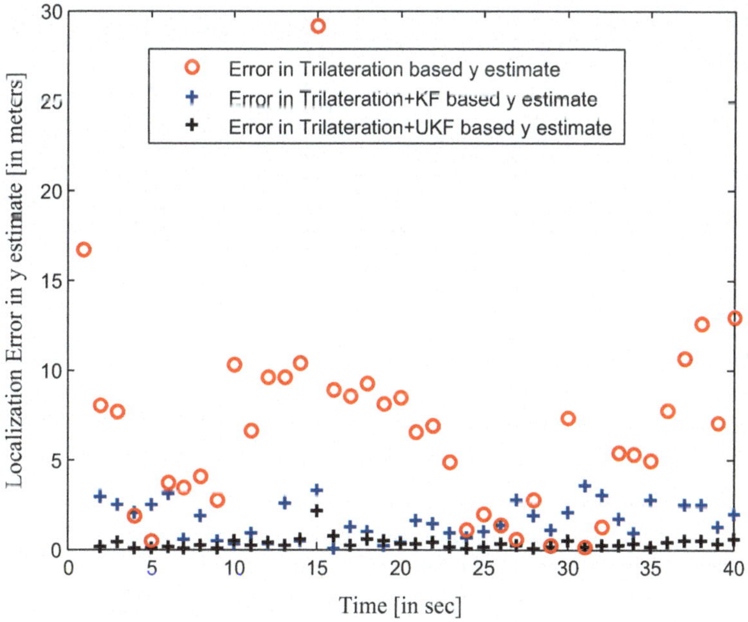

Fig. 5.9 Target localization error in y estimate with trilateration, trilateration+KF, and trilateration+UKF algorithms (Case I)

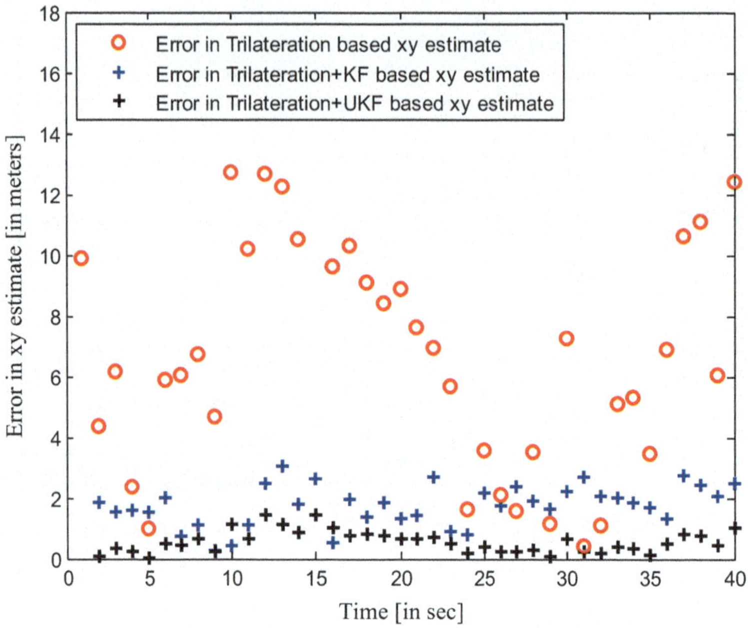

Fig. 5.10 Target localization error in x–y estimate with trilateration, trilateration+KF, and trilateration+UKF algorithms (Case I)

Table 5.6 Anchor node deployment in Case II

Anchor node	2-D location (in meters)
1	(0, 0)
2	(100, 0)
3	(0, 100)
4	(100, 100)
5	(30, 40)
6	(30, 60)
7	(50, 80)
8	(70, 60)

In Case II, the target starts moving from location (10, 20) and stops at (90, 90) as shown in Fig. 5.11. These figures depict the actual and the estimated target tracks using the trilateration, trilateration+KF, and trilateration+UKF algorithms. The color combinations for various terms in Fig. 5.11 are as given below:

For anchor nodes: black filled squares.
For actual target location at any time instance k: red unfilled squares.
For estimated location using trilateration: red plus symbol.
For estimated location using trilateration + KF: blue plus symbol.
For estimated location using trilateration + UKF: black circle symbol.

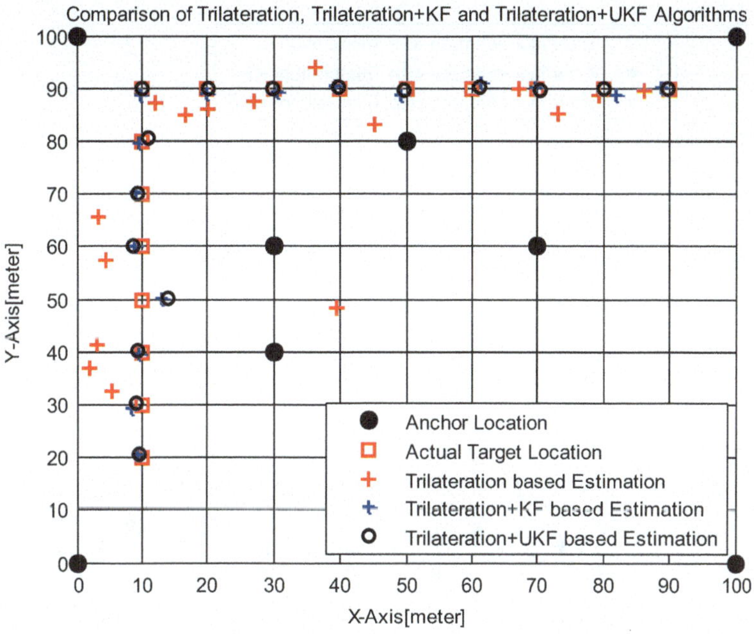

Fig. 5.11 Actual and estimated target tracks using trilateration, trilateration+KF, and trilateration+UKF algorithms (Case II)

Table 5.7 Numeric results obtained using trilateration, trilateration+KF, and trilateration+UKF target L&T algorithms (Case II)

Case II, 3 dBm, 100 × 100 area			
Sr no	Measurement noise (dBm)	Avg localization error (in meters)	RMSE
1	Trilateration	4.7318	9.0655
2	Trilateration+KF	0.9499	1.6527
3	Trilateration+UKF	0.4058	1.1339

The numeric results in Case II are presented in Table 5.7. From these results it is clear that the average localization error and RMSE is highest for trilateration-based estimation, moderate for trilateration+KF-based estimation, and lowest for trilateration+UKF-based estimation. That means the trilateration+UKF-based target L&T algorithm is superior in L&T performance than the rest of the two L&T algorithms. The % reductions in RMSE in the trilateration+UKF-based L&T algorithm as compared to the trilateration+KF algorithm and trilateration based L&T algorithm are 31% and 87%, respectively. Thus, we can easily claim that KF and UKF algorithms are able to improve the target L&T performance of the trilateration-based estimation technique further. In other words, the proposed trilateration+KF and trilateration+UKF algorithms can better deal with drastic variations in the velocity of the target and the environmental dynamicity (Figs. 5.12, 5.13, and 5.14).

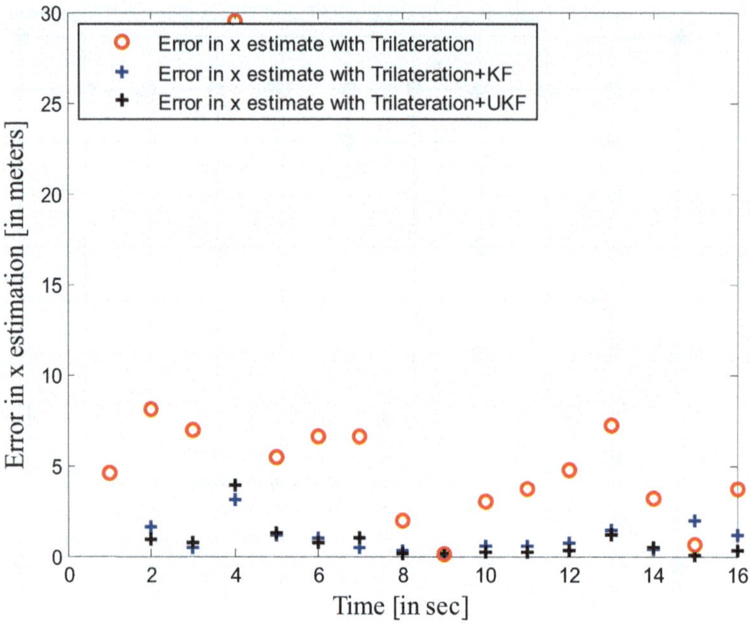

Fig. 5.12 Target localization error in *x* estimate with trilateration, trilateration+KF, and trilateration+UKF algorithms (Case II)

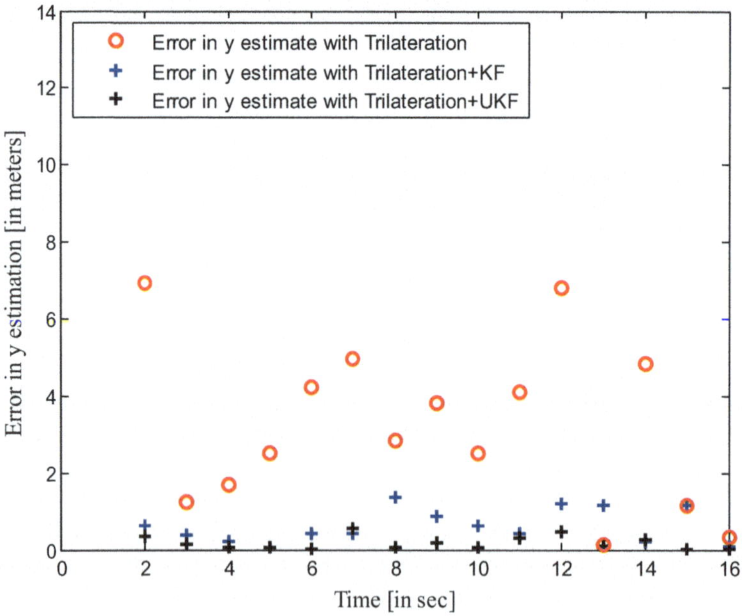

Fig. 5.13 Target localization error in *y* estimate with trilateration, trilateration+KF, and trilateration+UKF algorithms (Case II)

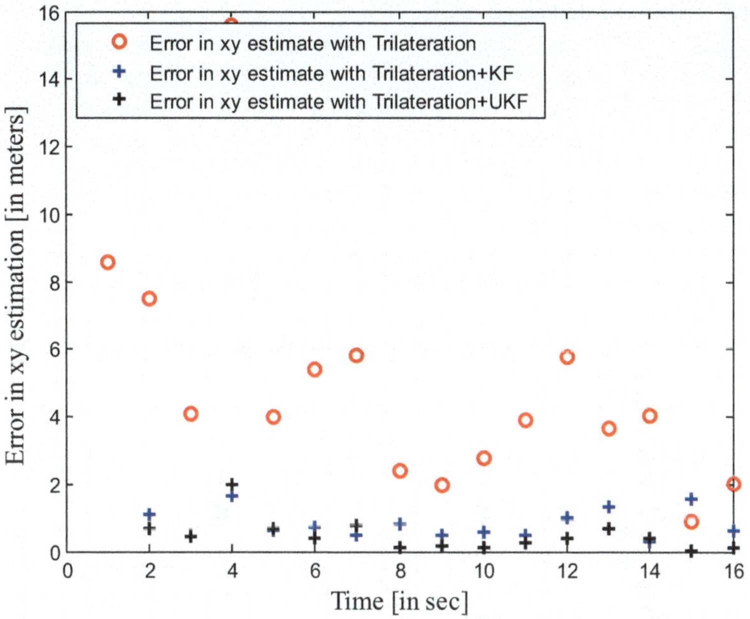

Fig. 5.14 Target localization error in *x–y* estimate with trilateration, trilateration+KF, and trilateration+UKF algorithms (Case II)

5.4.3 Case III Results

Testing the proposed trilateration+KF and trilateration+UKF L&T algorithms for linear target trajectory with anchor density = 8 and WSN area of 200 m × 200 m.

In Case III, the anchor nodes are kept at locations in WSN-defined area as given in Table 5.8. In Case III, the target starts moving from location (10, 20) and stops at (90, 90) as shown in Fig. 5.15. These figures depict the actual and the estimated target tracks using the trilateration, trilateration+KF, and trilateration+UKF algorithms. The color combinations for various terms in Fig. 5.15 are as given below:

For anchor nodes: black filled squares.
For actual target location at any time instance k: red unfilled squares.
For estimated location using trilateration: red plus symbol.
For estimated location using trilateration + KF: blue plus symbol.
For estimated location using trilateration+UKF: black circle symbol.

The numeric results in Case III are presented in Table 5.9. From these results it is clear that the average localization error and RMSE is highest for trilateration-based estimation, moderate for trilateration+KF-based estimation, and lowest for trilateration+UKF-based estimation. That means the trilateration+UKF-based target L&T algorithm is superior in L&T performance than the rest of the two L&T algorithms. The % reductions in RMSE in the trilateration+UKF-based L&T algorithm as compared to the trilateration+KF algorithm and trilateration algorithm are 9%

Table 5.8 Anchor node deployment in Case III

Anchor node	2-D location	Anchor node	2-D location
1	(0, 0)	5	(60, 80)
2	(200, 0)	6	(60, 120)
3	(0, 200)	7	(100, 160)
4	(200, 200)	8	(140, 120)

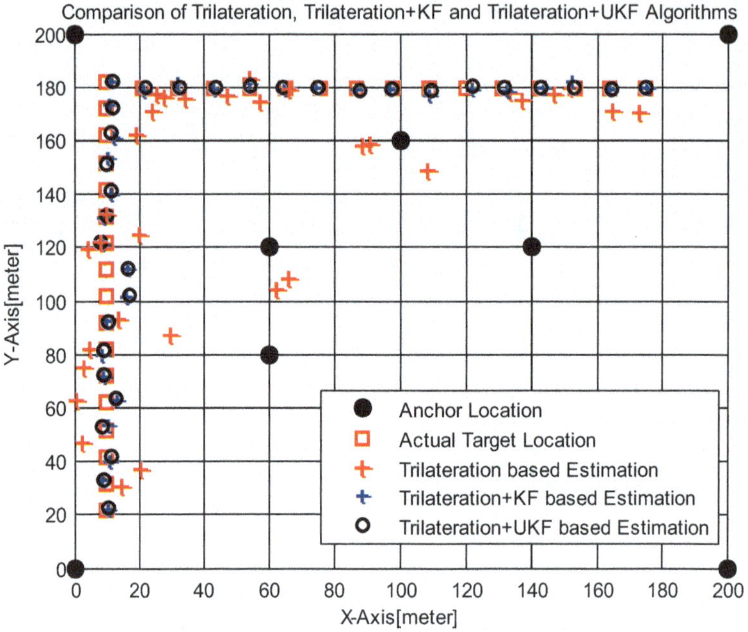

Fig. 5.15 Actual and estimated target tracks using trilateration, trilateration+KF, and trilateration+UKF algorithms (Case III)

Table 5.9 Numeric results obtained using trilateration, trilateration+KF, and trilateration+UKF target L&T algorithms (Case II)

Case III, 3 dBm, 200 × 200 area			
Sr no	Measurement noise (dBm)	Avg localization error (in meters)	RMSE
1	Trilateration	8.8301	19.4035
2	Trilateration+KF	1.0963	2.2284
3	Trilateration+UKF	0.8415	2.0145

and 88%, respectively. Thus, we can easily claim that KF and UKF algorithms are able to improve the target L&T performance of the trilateration-based estimation technique further. In other words, the proposed trilateration+KF and trilateration+UKF algorithms can better deal with the abrupt changes in the target velocity as well as the environmental dynamicity (Figs. 5.16, 5.17, and 5.18).

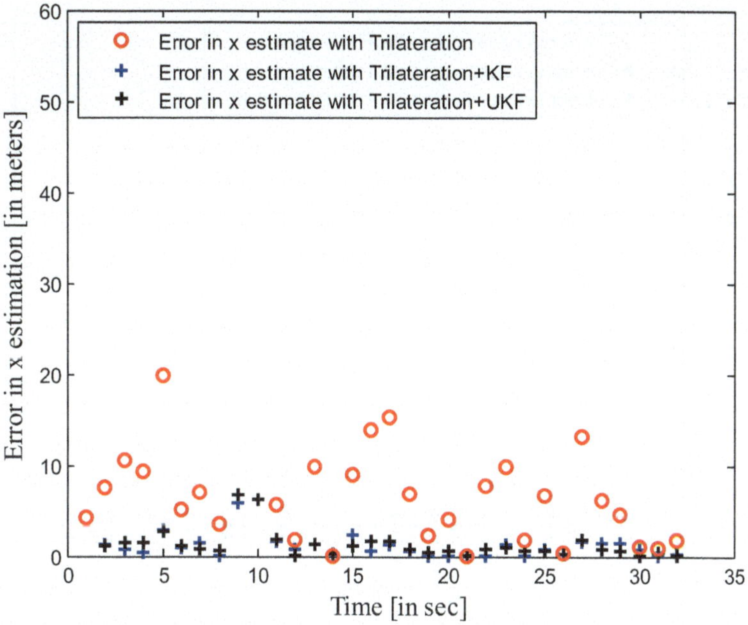

Fig. 5.16 Target localization error in *x* estimate with trilateration, trilateration+KF, and trilateration+UKF algorithms (Case III)

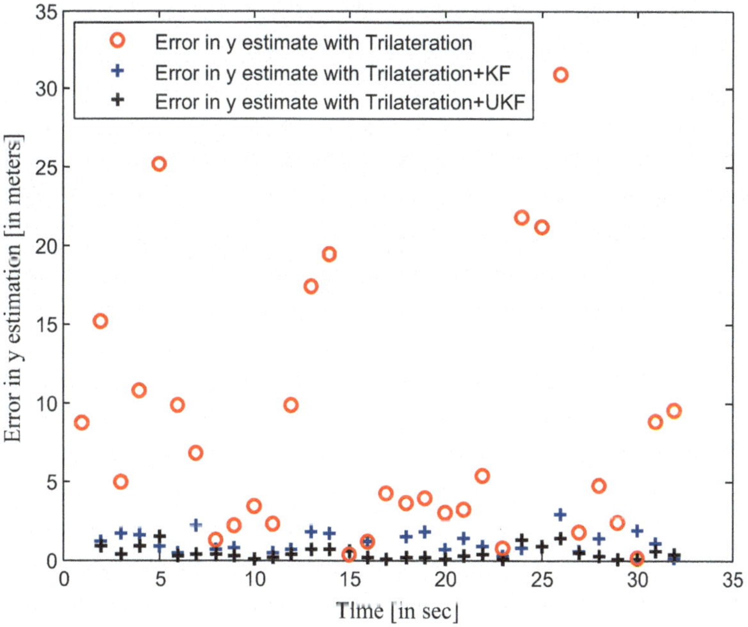

Fig. 5.17 Target localization error in *y* estimate with trilateration, trilateration+KF, and trilateration+UKF algorithms (Case III)

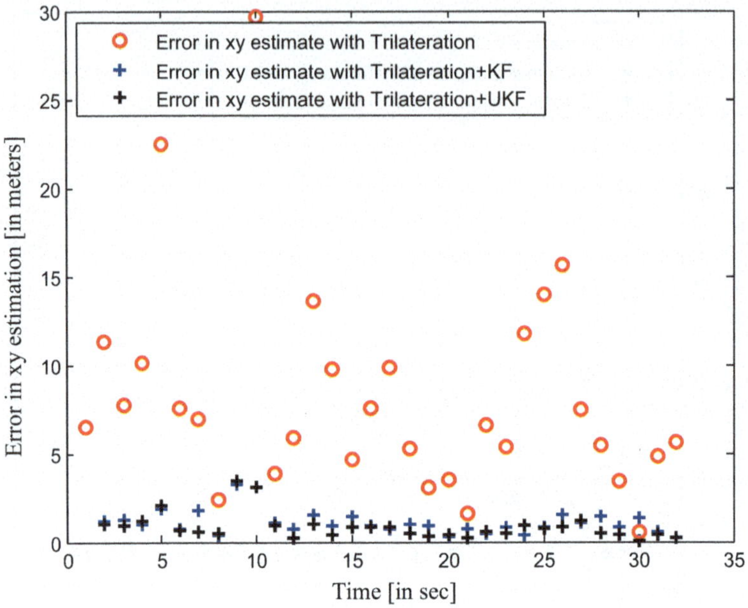

Fig. 5.18 Target localization error in *x*–*y* estimate with trilateration, trilateration+KF, and trilateration+UKF algorithms (Case III)

5.5 Conclusions

This chapter discusses the application of KF framework to the problem of L&T of the mobile target using WSN with uncertainty in RSSI measurement noise. The target trajectories are kept nonlinear (Case I) and linear (Case II and Case III) to understand the impact of the changes in the target trajectories on the proposed target L&T algorithms. The extensive simulation experiments carried out demonstrate high target tracking accuracy, irrespective of the environmental dynamicity and abrupt changes in the target velocity. There are many parameters that impact the performance of RSSI-based L&T algorithm, namely, drastic changes in the target velocity and measurement noise in the given RF environment. To understand the impact of drastic variations in the target velocity, we varied the target velocity abruptly in the range of −2 to 7 m/s in Case I, 0 to 10 m/s in Case II, and 0 to 12 m/s in Case III at specific time instances. To study the impact of the environmental dynamicity (fluctuations), the measurement noise in the RSSI is kept constant at 3 dBm. The simulation results prove that the trilateration+UKF-based target L&T algorithm outperforms the trilateration and trilateration+KF-based target L&T algorithms, irrespective of drastic variations in the target velocity, changes in the target trajectories, as well as the environmental dynamicity. Thus, the KF framework is capable to further refine the location estimates given by trilateration.

MATLAB Code for KF-Based Target L&T

Main_File:

```
%%%%%%%%%%%%%%%%%%%%%%%%%%%%%%%%Main_file.m%%%%%%%%%%%%%%%%%%%%%%%%%%%%%

%% WSN Deployment Setting Parameters
    clear all
    close all
    clc

    networkSize = 100;  % we consider a 100by100 area in which target moves

     anchorLoc    = [0                      0;
                     networkSize/2          networkSize*0.2;
                     networkSize            0;
                     networkSize            networkSize;
                     networkSize/2          networkSize*0.8;
                     0                      networkSize];


%show anchor Locations
    f1 = figure(1);

plot(anchorLoc(:,1),anchorLoc(:,2),'ko','MarkerSize',8,'lineWidth',2,'MarkerF
     aceColor','k');

    grid on
    hold on

% Defining veriables
    no_of_positions = 40;


P = [0.25 0 0  0; 0 0.04 0 0; 0 0 0.02 0; 0 0 0 0.01];
% Initial Process Covariance Matrix
    q=0.1;                                  %std of process


error_x_kf = zeros( no_of_positions,1);
error_y_kf = zeros(no_of_positions,1);
error_xy_kf = zeros(no_of_positions,1);
error_x_ukf = zeros(no_of_positions,1);
error_y_ukf = zeros(no_of_positions,1);
error_xy_ukf = zeros(no_of_positions,1);
error_x_rssi = zeros(no_of_positions,1);
error_y_rssi = zeros(no_of_positions,1);
error_xy_rssi = zeros(no_of_positions,1);

    RMSE_kf =0 ; RMSE_rssi= 0; RMSE_ukf= 0;

% Calculate reference RSSI at d0 = 1 meter using Free Space Path Loss Model
    d0=1;                                                    % in Meters
    Pr0 = RSSI_friss(d0);

    d_test = 20;

    Pr = RSSI_friss(d_test);

%Calculation of Path Loss Exponent
      n = -(Pr + Pr0)/(10*log(d_test))

      x=10;
      y=10;
```

```
% Generating trajectory for the mobile node
for t = 1:no_of_positions
if(t<9)                     x_v = 3; y_v = 7;
elseif(t==9 && t<16)        x_v = 4; y_v = 2;
elseif(t>=16 && t<18)       x_v = 0; y_v = 0;
elseif(t>=18 && t<30)       x_v = 2; y_v = -3;
elseif(t>=30 && t<=40)      x_v = -2; y_v = -2;
end

       x=x+x_v;
       y=y+y_v;

       disp('True Location')
       [x,y]                              % Actual Target Location
       plot(x,y,'rs','LineWidth',2)

       title('Evaluation of Trilateration, Trilateration+KF &
              Trilateration+UKF','FontName','Times','FontSize',12);
       ylabel('Y-Axis[meter]','FontName','Times','Fontsize',14);
       xlabel('X-Axis[meter]','FontName','Times','Fontsize',14);

       legend('Anchor Location','Actual Target Location','Trilateration
               based Estimation','Trilateration+KF based Estimation',
'Trilateration+UKF based Estimation','Location','SouthEast')
       hold on

% Actual Distances from Anchors required to generate RSSI Values

       d1 = sqrt( x^2 + y^2 );
       d2 = sqrt((networkSize/2 - x)^2 + (networkSize*0.2 - y)^2);
       d3 = sqrt((networkSize - x)^2 + y^2);
       d4 = sqrt((networkSize - x)^2+ (networkSize - y)^2);
       d5 = sqrt((networkSize/2 - x)^2+ (networkSize*0.8 - y)^2);
       d6 = sqrt(x^2+ (networkSize - y)^2);


% Generate RSSI Values at 6 Anchor Nodes which are at d1, d2, d3, d4, d5 &
d6 distances respectively from Moving Target

       RSS = lognormalshadowing_6(n,d1,d2,d3,d4,d5,d6,Pr0);
       RSS_s = sort(RSS);
end

     disp('Trilateration based Estimated Location')

       mobileLoc_est = trilateration_6(RSS,RSS_s,Pr0,n,networkSize)
       X_T = mobileLoc_est(1);
       Y_T = mobileLoc_est(2);

       plot(X_T,Y_T,'r+','LineWidth',2,'MarkerSize',8);
       hold on

% calculate velocities in X & Y Directions

          velocity_est=velocity(X_T,Y_T,t);

          s = [x; y; 0; 0];               %State of the system at time 't'
X = s + q*randn(4,1);          % state with noise
```

```
% KF for Tracking Moving Target Code starts here
       [X_kalman,Y_kalman,X_kf]= kf(X,P,X_T,Y_T,velocity_est,t);
       plot(X_kalman,Y_kalman,'b+','LineWidth',2)
       hold on

%UKF implementation
      Z = [X_T; Y_T; velocity_est(1); velocity_est(2)];
      disp('UKF Estimated State')
      [X_ukf]= ukf5(X,P,Z)
      plot(X_ukf(1),X_ukf(2),'ko','LineWidth',2)
      hold on

% Error Analysis of algorithm
% ---> Part 1: for Pure RSSI based Technique
       RMSE_rssi = RMSE_rssi + ((X_T - x)^2 + (Y_T - y)^2);

       RMSE_kf = RMSE_kf + ((X_kalman - x)^2 + (Y_kalman - y)^2);
       RMSE_ukf = RMSE_ukf + ((X_ukf(1)-x)^2 + (X_ukf(2)-y)^2);

% ---> Part 2: Calculation of Absolute Errors

% a) For Kalman Filter
       error_x_kf(t) =  abs((x - X_kalman));
       error_y_kf(t) =  abs((y - Y_kalman));
       error_xy_kf(t) = ((error_x_kf(t) + error_y_kf(t))/2);

% b) For Unscented Kalman Filter
       error_x_ukf(t) = abs((x - X_ukf(1)));
       error_y_ukf(t) = abs((y - X_ukf(2)));
       error_xy_ukf(t) = ((error_x_ukf(t) + error_y_ukf(t))/2);

% c) For Trilateration based Technique

       error_x_rssi(t) = abs((x - X_T));
       error_y_rssi(t) = abs((y - Y_T));
       error_xy_rssi(t) =((error_x_rssi(t) + error_y_rssi(t))/2);

end

% Average Error in x & y  coordinates
      avg_error_xy_rssi = 0;
      avg_error_xy_kf = 0 ;
      avg_error_xy_ukf = 0 ;

for t = 1:no_of_positions

avg_error_xy_rssi=avg_error_xy_rssi + (error_xy_rssi(t)/no_of_positions);
avg_error_xy_kf= avg_error_xy_kf + (error_xy_kf(t)/no_of_positions);
avg_error_xy_ukf= avg_error_xy_ukf + (error_xy_ukf(t)/no_of_positions);

end
      avg_error_xy_rssi
      avg_error_xy_kf
      avg_error_xy_ukf
```

```
% Average Error in x & y  coordinates
      RMSE_rssi = sqrt(RMSE_rssi/no_of_positions)
      RMSE_kf = sqrt(RMSE_kf/no_of_positions)
      RMSE_ukf = sqrt(RMSE_ukf/no_of_positions)


% Plotting Absolute Errors of KF & UKF based Tracking

f2 = figure(2);
for t =1:no_of_positions

    plot(t,error_x_kf(t),'b+','Linewidth',2)
plot(t,error_x_ukf(t),'k+','Linewidth',2)
plot(t,error_x_rssi(t),'ro','Linewidth',2)

xlabel('Time [in sec]','FontName','Times','Fontsize',14)
ylabel('Localization Error in x estimate [in meters]','FontName','Times',
'Fontsize',14)
    hold on

end

  legend('Error in Trilateration based x estimate','Error in
Trilateration+KF based x estimate','Error in Trilateration+UKF
based x estimate','Location','NorthWest')

  f3 = figure(3);
for t =1:no_of_positions

    plot(t,error_y_kf(t),'b+','Linewidth',2)
    plot(t,error_y_ukf(t),'k+','Linewidth',2)
    plot(t,error_y_rssi(t),'ro','Linewidth',2)
    xlabel('Time [in sec]','FontName','Times','Fontsize',14)
    ylabel('Localization Error in y estimate [in meters]','FontName',
 'Times','Fontsize',14)

  hold on

end

legend('Error in Trilateration based y estimate','Error in Trilateration+KF
based y estimate','Error in Trilateration+UKF based y estimate',
'Location','NorthWest')

  f4 = figure(4);
 for t =1:no_of_positions

plot(t,error_xy_kf(t),'b+','Linewidth',2)
plot(t,error_xy_ukf(t),'k+','Linewidth',2)
plot(t,error_xy_rssi(t),'ro','Linewidth',2)

 xlabel('Time [in sec]','FontName','Times','Fontsize',14)
 ylabel('Error in xy estimate [in meters]','FontName','Times','Fontsize',14)

  hold on

 end
```

```
legend('Error in Trilateration based xy estimate','Error in Trilateration+KF
based xy estimate','Error in Trilateration+UKF based xy estimate',
'Location','NorthWest')
```

%%% kf.m %%%%%%%%%%%%%%%%%%%%%%%%%%%%

```
function [ X_kalman,Y_kalman, X_kf] = kf( X,P,X_T,Y_T,velocity_est,t)
%UNTITLED2 Summary of this function goes here
%   Detailed explanation goes here

% Kalman Filter Configuration Parameters
       x_ini = 10; y_ini = 20;
       x_v = 0; y_v = 0;    % velocities
       dt = 1;  % time interval of 1 second

       A = [1 0 dt 0; 0 1 0 dt; 0 0 1 0; 0 0 0 1];
       B = [(1/2)*dt^2  0; 0 (1/2)*dt^2; dt 0; 0 dt];
       dt=1;
       U = [3.5; 3.5];              % Control Input (acceleration)

var_x = 2; var_y = 2;
var_x_v = 0; var_y_v = 0;     % Variance in process

       R = [2.2 0 0 0     ; 0 1.2 0 0; 0 0 0.9 0; 0 0 0 0.5];
       sig = 1;

Q  =  sig*[(1/3)*dt^3   0    (1/2)*dt^2    0;  0    (1/3)*dt^3   0     (1/2)*dt^2;
(1/2)*dt^2  0  dt  0;  0  (1/2)*dt^2  0  dt];

       H = eye(4);
       I = eye(4);

% Prediction Stage
       X = A*X + B*U + [randn;randn;randn;randn];
       P = A*P*A'; + Q; % Q process noise covariance matrix

% Update Stage
       X_meas = [X_T; Y_T; velocity_est(1); velocity_est(2)];
% Observation Equation

       Y = H*X_meas + [randn;randn;randn;randn];
       K = (P*H)/((H*P*H')+ R);
       disp('Kalman Filter State Location')
       X = X + K*(Y-(H*X))

       X_kf = X;
       X_kalman=X(1);
       Y_kalman=X(2);
       P = (I - (K*H))*P;

end
```

%%%% ukf.m %%%%%%%%%%%%%%%%%%%%%%%%%%%%

```
function [ X_ukf ] = ukf5( X,P,Z )
%UNTITLED4 Summary of this function goes here
%   Detailed explanation goes here

   dt= 1;
   A = [1 0 dt 0; 0 1 0 dt; 0 0 1 0; 0 0 0 1];
   H = eye(4);
   R = [2.2 0 0 0     ; 0 1.2 0 0; 0 0 0.9 0; 0 0 0 0.5];
```

```
  %R = [0.00001 0 0 0;   0  0.00001 0 0;  0  0 0.00001 0; 0 0 0 0.00001];
  %Q = eye(4);
      dt=1;
    sig = 1;

Q  =  sig*[(1/3)*dt^3   0    (1/2)*dt^2    0;  0    (1/3)*dt^3   0    (1/2)*dt^2;
(1/2)*dt^2  0  dt  0;  0  (1/2)*dt^2  0  dt];


    L=numel(X);                                  %numer of states
    m=numel(Z);                                  %numer of measurements
    alpha=1e-3;                                  %default, tunable
    ki=0;                                        %default, tunable
    beta=2;                                      %default, tunable
    lambda=alpha^2*(L+ki)-L;                     %scaling factor
    c=L+lambda;                                  %scaling factor

   Wm=[lambda/c 0.5/c+zeros(1,2*L)];             %weights for means
    Wc=Wm;
    Wc(1)=Wc(1)+(1-alpha^2+beta);                  %weights for covariance
    c=sqrt(c);

  % Sigma Point Calculation
  %X=sigmas(x,P,c);                              %sigma points around x

  D = c*chol(P)';
    Y = X(:,ones(1,numel(X)));
    X = [X Y+D Y-D];

    [x1,X1,P1,X2]=ut(A,X,Wm,Wc,L,Q);  %unscented transformation of process
    [z1,Z1,P2,Z2]=ut(H,X1,Wm,Wc,m,R); %unscented transformation of measurments

    P12=X2*diag(Wc)*Z2';                          %transformed cross-covariance
    K=P12*inv(P2);
    X=x1+K*(Z-z1);                                %state update
    X_ukf = X;
    P=P1-K*P12';                                  %covariance update

  end
```

```
  %% WSN Deployment Setting Parameters
      clear all
      close all
      clc

      networkSize = 100;

       anchorLoc    = [0                     0;
                       networkSize*0.3       networkSize*0.4;
                       networkSize           0;
                       networkSize*0.7       networkSize*0.6;
                       networkSize           networkSize;
                       networkSize/2         networkSize*0.8;
                       0                     networkSize;
                       networkSize*0.3       networkSize*0.6];

  %show anchor Locations
      f1 = figure(1);

plot(anchorLoc(:,1),anchorLoc(:,2),'ko','MarkerSize',8,'lineWidth',2,'MarkerF
aceColor','k');
      grid on
      hold on

  % Defining veriables
      no_of_positions = 16;

      P = [0.25 0 0  0; 0 0.04 0 0; 0 0 0.02 0; 0 0 0 0.01];    % Initial
Process Covariance Matrix

  error_x_kf = zeros( no_of_positions,1);
  error_y_kf = zeros(no_of_positions,1);
  error_xy_kf = zeros(no_of_positions,1);
  error_x_ukf = zeros(no_of_positions,1);
  error_y_ukf = zeros(no_of_positions,1);
  error_xy_ukf = zeros(no_of_positions,1);
  error_x_rssi = zeros(no_of_positions,1);
  error_y_rssi = zeros(no_of_positions,1);
  error_xy_rssi = zeros(no_of_positions,1);


      RMSE_kf =0 ; RMSE_rssi= 0; RMSE_ukf= 0; RMSE_kf_V= 0;

  % Calculate reference RSSI at d0 = 1 meter using Free Space Path Loss Model
      d0=1;                                                        % in Meters
      Pr0 = RSSI_friss(d0);

      d_test = 20;

      Pr = RSSI_friss(d_test);

  %Calculation of Path Loss Exponent:
```

```
        n = -(Pr + Pr0)/(10*log(d_test));

  % Initial Position of Target
        x=10;
        y=10;

  % Generating trajectory for the mobile node
  for t = 1:no_of_positions
  if(t<9)
         x=10;
         y=x*t+10;
  end

  if(t>=9)
         y=90;
         x=x+10;
  end

         disp('True Location');

         mobileLoc1 = [x,y]                          % Actual Target Location
         plot(x,y,'rs','LineWidth',2,'MarkerSize',8)
         ylabel('Y-Axis[meter]');
         xlabel('X-Axis[meter]');

         title('Comparison of Trilateration, Trilateration+KF and
                Trilateration+UKF Algorithms');

legend('Anchor Location','Actual Target Location','Trilateration based
        Estimation','Trilateration+KF based Estimation','Trilateration+UKF
        based Estimation','Location','SouthEast')
         hold on

  % Actual Distances from Anchors required to generate RSSI Values

         d1 = sqrt( x^2 + y^2 );
         d2 = sqrt((networkSize*0.3 - x)^2 + (networkSize*0.4 - y)^2);
         d3 = sqrt((networkSize - x)^2 + y^2);
         d4 = sqrt((networkSize*0.7 - x)^2+ (networkSize*0.6 - y)^2);
         d5 = sqrt((networkSize - x)^2+ (networkSize - y)^2);
         d6 = sqrt((networkSize/2 - x)^2+ (networkSize*0.8 - y)^2)
         d7 = sqrt(x^2+ (networkSize - y)^2)
         d8 = sqrt((networkSize*0.3 - x)^2 + (networkSize*0.6 - y)^2)


  % Generate RSSI Values
         RSS = lognormalshadowing_8(n,d1,d2,d3,d4,d5,d6,d7,d8,Pr0);
         RSS_s = sort(RSS);

         disp('Estimated Location')

         mobileLoc_est = trilateration_8(RSS,RSS_s,Pr0,n,networkSize)
         X_T = mobileLoc_est(1);
         Y_T = mobileLoc_est(2);
```

```
        plot(X_T,Y_T,'r+','LineWidth',2,'MarkerSize',8);
        hold on

% calculate velocities in X & Y Directions

        velocity_est=velocity(X_T,Y_T,t);
           X = [x; y; 0; 0];                % State of the system at time 't'

% Kalman Filter for Tracking Moving Target Code starts here

[X_kalman,Y_kalman,X_V_kalman, Y_V_kalman]= kf(X,P,X_T,Y_T,velocity_est,t);

        plot(X_kalman,Y_kalman,'b+','LineWidth',2)
        hold on

%UKF implementation
       Z = [X_T; Y_T; velocity_est(1); velocity_est(2)];
       disp('UKF Estimated State');
       [X_ukf]= ukf5(X,P,Z)

       plot(X_ukf(1),X_ukf(2),'ko','LineWidth',2)
       hold on

% Error Analysis of algorithm
% ---> Part 1 : for Pure RSSI based Technique
        RMSE_rssi = abs(RMSE_rssi + ((X_T - x)^2 + (Y_T - y)^2));

        RMSE_kf = RMSE_kf + ((X_kalman - x)^2 + (Y_kalman - y)^2);
        RMSE_ukf = RMSE_ukf + ((X_ukf(1)-x)^2 + (X_ukf(2)-y)^2);

% ---> Part 2 : Calculation of Absolute Errors

% a) For Kalman Filter
        error_x_kf(t) =  abs((x - X_kalman));
        error_y_kf(t) =  abs((y - Y_kalman));
        error_xy_kf(t) = ((error_x_kf(t) + error_y_kf(t))/2);

% b) For Unscented Kalman Filter
        error_x_ukf(t) = abs((x - X_ukf(1)));
        error_y_ukf(t) = abs((y - X_ukf(2)));
        error_xy_ukf(t) = ((error_x_ukf(t) + error_y_ukf(t))/2);

% c) For Pure RSSI based Technique
        error_x_rssi(t) = abs((x - X_T));
        error_y_rssi(t) = abs((y - Y_T));
        error_xy_rssi(t) =((error_x_rssi(t) + error_y_rssi(t))/2);

end

% Average Error in x & y  coordinates
       avg_error_xy_rssi = 0;
       avg_error_xy_kf = 0 ;
       avg_error_xy_ukf = 0 ;
```

```
for t = 1:no_of_positions

avg_error_xy_rssi=avg_error_xy_rssi + (error_xy_rssi(t)/no_of_positions);
avg_error_xy_kf= avg_error_xy_kf + (error_xy_kf(t)/no_of_positions);
avg_error_xy_ukf= avg_error_xy_ukf + (error_xy_ukf(t)/no_of_positions);

end

     avg_error_xy_rssi
     avg_error_xy_kf
     avg_error_xy_ukf

      RMSE_rssi = sqrt(RMSE_rssi/no_of_positions)
      RMSE_kf = sqrt(RMSE_kf/no_of_positions)
      RMSE_ukf = sqrt(RMSE_ukf/no_of_positions)


% Plotting Absolute Errors of KF & UKF based Tracking

f2 = figure(2);
for t =1:no_of_positions

    plot(t,error_x_kf(t),'b+','Linewidth',2)
    plot(t,error_x_ukf(t),'k+','Linewidth',2)
    plot(t,error_x_rssi(t),'ro','Linewidth',2)

xlabel('Time [in sec]','FontName','Times','Fontsize',12)
ylabel('Error in x estimation [in meters]','FontName','Times','Fontsize',12)

hold on

end

legend('Error in x estimate with Trilateration','Error in x estimate with
        Trilateration+KF','Error in x estimate with Trilateration+UKF',
'Location','NorthWest')

  f3 = figure(3);
for t =1:no_of_positions

    plot(t,error_y_kf(t),'b+','Linewidth',2)
    plot(t,error_y_ukf(t),'k+','Linewidth',2)
    plot(t,error_y_rssi(t),'ro','Linewidth',2)

 xlabel('Time [in sec]','FontName','Times','Fontsize',12)
 ylabel('Error in y estimation [in meters]','FontName', 'Times',
'Fontsize',12)

  hold on

end
legend('Error in y estimate with Trilateration','Error in y estimate with
          Trilateration+KF','Error in y estimate with Trilateration+UKF',
```

```
'Location','NorthWest')

    f4 = figure(4);
  for t =1:no_of_positions

      plot(t,error_xy_kf(t),'b+','Linewidth',2)
      plot(t,error_xy_ukf(t),'k+','Linewidth',2)
      plot(t,error_xy_rssi(t),'ro','Linewidth',2)

   xlabel('Time [in sec]','FontName','Times','Fontsize',12)
   ylabel('Error in xy estimation [in meters]','FontName'
  ,'Times','Fontsize',12)

    hold on
  end

legend('Error in xy estimate with Trilateration','Error in xy estimate with
        Trilateration+KF','Error in xy estimate with
        Trilateration+UKF','Location','NorthWest')

  %linearity_test(n,Pr0);

  f6 = figure(6);
  for t =1:no_of_positions

  if(t<=8)
            x_v=10;
            plot(t,x_v,'b+','Linewidth',2)
  elseif(t>8)
            x_v=0;
            plot(t,x_v,'b+','Linewidth',2)
  end
        hold on

   xlabel('Time, [s]','FontName','Times','Fontsize',12)
   ylabel('Target Velocity in X direction [m/sec]','FontName',
  'Times','Fontsize',12)
  end

    f7 = figure(7);
  for t =1:no_of_positions

  if(t<=8)
            y_v=0;
            plot(t,y_v,'b+','Linewidth',2)
  elseif(t>8)
            y_v=10;
            plot(t,y_v,'b+','Linewidth',2)
  end
        hold on
xlabel('Time, [s]','FontName','Times','Fontsize',12)
ylabel('Target Velocity in Y direction [m/sec]','FontName',
'Times','Fontsize',12)
  end
```

```
%% WSN Deployment Setting Parameters
    clear all
    close all
    clc

    networkSize = 200;
    anchorLoc   = [0                    0;
                   networkSize*0.3      networkSize*0.4;
                   networkSize          0;
                   networkSize*0.7      networkSize*0.6;
                   networkSize          networkSize;
                   networkSize/2        networkSize*0.8;
                   0                    networkSize;
                   networkSize*0.3      networkSize*0.6];

%show anchor Locations
    f1 = figure(1);

plot(anchorLoc(:,1),anchorLoc(:,2),'ko','MarkerSize',8,'lineWidth',2,'MarkerF
aceColor','k');
    grid on
    hold on

% Defining veriables
    no_of_positions = 32;

    P = [0.25 0 0  0; 0 0.04 0 0; 0 0 0.02 0; 0 0 0 0.01];

error_x_kf = zeros( no_of_positions,1);
error_y_kf = zeros(no_of_positions,1);
error_xy_kf = zeros(no_of_positions,1);
error_x_ukf = zeros(no_of_positions,1);
error_y_ukf = zeros(no_of_positions,1);
error_xy_ukf = zeros(no_of_positions,1);
error_x_rssi = zeros(no_of_positions,1);
error_y_rssi = zeros(no_of_positions,1);
error_xy_rssi = zeros(no_of_positions,1);


    RMSE_kf =0 ; RMSE_rssi= 0; RMSE_ukf= 0; RMSE_kf_V= 0;

% Calculate reference RSSI at d0 = 1 meter using Free Space Path Loss Model
    d0=1;                                                       % in Meters
    Pr0 = RSSI_friss(d0);

    d_test = 20;

    Pr = RSSI_friss(d_test);

      n = -(Pr + Pr0)/(10*log(d_test));

% Initial Position of Target
```

```
        x=10;
        y=10;

  % Generating trajectory for the mobile node
  for t = 1:no_of_positions
  if(t<18)
         x=10;
         y=x*t+12;
  end

  if(t>=18)
         y=180;
         x=x+11;
  end

         disp('True Location');
         mobileLoc1 = [x,y]                     % Actual Target Location
         plot(x,y,'rs','LineWidth',2,'MarkerSize',8)
         ylabel('Y-Axis[meter]');
         xlabel('X-Axis[meter]');

title('Comparison of Trilateration, Trilateration+KF and Trilateration+UKF
Algorithms');

legend('Anchor   Location','Actual   Target   Location','Trilateration   based
Estimation','Trilateration+KF   based   Estimation','Trilateration+UKF   based
Estimation','Location','SouthEast')
         hold on

  % Actual Distances from Anchors required to generate RSSI Values

         d1 = sqrt( x^2 + y^2 );
         d2 = sqrt((networkSize*0.3 - x)^2 + (networkSize*0.4 - y)^2);
         d3 = sqrt((networkSize - x)^2 + y^2);
         d4 = sqrt((networkSize*0.7 - x)^2+ (networkSize*0.6 - y)^2);
         d5 = sqrt((networkSize - x)^2+ (networkSize - y)^2);
         d6 = sqrt((networkSize/2 - x)^2+ (networkSize*0.8 - y)^2)
         d7 = sqrt(x^2+ (networkSize - y)^2)
         d8 = sqrt((networkSize*0.3 - x)^2 + (networkSize*0.6 - y)^2)


  % Generate RSSI Values
         RSS = lognormalshadowing_8(n,d1,d2,d3,d4,d5,d6,d7,d8,Pr0);
         RSS_s = sort(RSS);

         disp('Estimated Location')

         mobileLoc_est = trilateration_8(RSS,RSS_s,Pr0,n,networkSize)
         X_T = mobileLoc_est(1);
         Y_T = mobileLoc_est(2);

         plot(X_T,Y_T,'r+','LineWidth',2,'MarkerSize',8);
         hold on
```

```
% calculate velocities in X & Y Directions

       velocity_est=velocity(X_T,Y_T,t);
          X = [x; y; 0; 0];          % State of the system at time 't'

% Kalman Filter for Tracking Moving Target Code starts here

[X_kalman,Y_kalman,X_V_kalman, Y_V_kalman]= kf(X,P,X_T,Y_T,velocity_est,t);

       plot(X_kalman,Y_kalman,'b+','LineWidth',2)
       hold on

%UKF implementation
      Z = [X_T; Y_T; velocity_est(1); velocity_est(2)];
      disp('UKF Estimated State');
      [X_ukf]= ukf5(X,P,Z)

      plot(X_ukf(1),X_ukf(2),'ko','LineWidth',2)
      hold on

% Error Analysis of algorithm
% ---> Part 1 : for Pure RSSI based Technique
       RMSE_rssi = abs(RMSE_rssi + ((X_T - x)^2 + (Y_T - y)^2));

       RMSE_kf = RMSE_kf + ((X_kalman - x)^2 + (Y_kalman - y)^2);
       RMSE_ukf = RMSE_ukf + ((X_ukf(1)-x)^2 + (X_ukf(2)-y)^2);

% ---> Part 2 : Calculation of Absolute Errors

% a) For Kalman Filter
       error_x_kf(t) =  abs((x - X_kalman));
       error_y_kf(t) =  abs((y - Y_kalman));
       error_xy_kf(t) = ((error_x_kf(t) + error_y_kf(t))/2);

% b) For Unscented Kalman Filter
       error_x_ukf(t) = abs((x - X_ukf(1)));
       error_y_ukf(t) = abs((y - X_ukf(2)));
       error_xy_ukf(t) = ((error_x_ukf(t) + error_y_ukf(t))/2);

% c) For Pure RSSI based Technique
       error_x_rssi(t) = abs((x - X_T));
       error_y_rssi(t) = abs((y - Y_T));
       error_xy_rssi(t) =((error_x_rssi(t) + error_y_rssi(t))/2);

end

% Average Error in x & y  coordinates
      avg_error_xy_rssi = 0;
      avg_error_xy_kf = 0 ;
      avg_error_xy_ukf = 0 ;

for t = 1:no_of_positions

avg_error_xy_rssi=avg_error_xy_rssi + (error_xy_rssi(t)/no_of_positions);
```

```
  avg_error_xy_kf= avg_error_xy_kf + (error_xy_kf(t)/no_of_positions);
  avg_error_xy_ukf= avg_error_xy_ukf + (error_xy_ukf(t)/no_of_positions);

  end

       avg_error_xy_rssi
       avg_error_xy_kf
       avg_error_xy_ukf


        RMSE_rssi = sqrt(RMSE_rssi/no_of_positions)
        RMSE_kf = sqrt(RMSE_kf/no_of_positions)
        RMSE_ukf = sqrt(RMSE_ukf/no_of_positions)


  % Plotting Absolute Errors of KF & UKF based Tracking
  f2 = figure(2);
  for t =1:no_of_positions

      plot(t,error_x_kf(t),'b+','Linewidth',2)
      plot(t,error_x_ukf(t),'k+','Linewidth',2)
      plot(t,error_x_rssi(t),'ro','Linewidth',2)

  xlabel('Time [in sec]','FontName','Times','Fontsize',12)
  ylabel('Error in x estimation [in meters]','FontName','Times','Fontsize',12)

  hold on

  end

legend('Error in x estimate with Trilateration','Error in x estimate with
        Trilateration+KF','Error in x estimate with Trilateration+UKF',
'Location','NorthWest')

    f3 = figure(3);
  for t =1:no_of_positions

      plot(t,error_y_kf(t),'b+','Linewidth',2)
      plot(t,error_y_ukf(t),'k+','Linewidth',2)
      plot(t,error_y_rssi(t),'ro','Linewidth',2)

   xlabel('Time [in sec]','FontName','Times','Fontsize',12)
   ylabel('Error in y estimation [in meters]','FontName', 'Times',
  'Fontsize',12)

    hold on

  end
legend('Error in y estimate with Trilateration','Error in y estimate with
          Trilateration+KF','Error in y estimate with Trilateration+UKF',
'Location','NorthWest')

    f4 = figure(4);
  for t =1:no_of_positions
```

```
      plot(t,error_xy_kf(t),'b+','Linewidth',2)
      plot(t,error_xy_ukf(t),'k+','Linewidth',2)
      plot(t,error_xy_rssi(t),'ro','Linewidth',2)

   xlabel('Time [in sec]','FontName','Times','Fontsize',12)
   ylabel('Error in xy estimation [in meters]','FontName'
  ,'Times','Fontsize',12)

    hold on
  end

legend('Error in xy estimate with Trilateration','Error in xy estimate with
        Trilateration+KF','Error in xy estimate with
        Trilateration+UKF','Location','NorthWest')

  %linearity_test(n,Pr0);

  f6 = figure(6);
  for t =1:no_of_positions

  if(t<=8)
            x_v=10;
            plot(t,x_v,'b+','Linewidth',2)
  elseif(t>8)
            x_v=0;
            plot(t,x_v,'b+','Linewidth',2)
  end
        hold on

   xlabel('Time, [s]','FontName','Times','Fontsize',12)
   ylabel('Target Velocity in X direction [m/sec]','FontName',
  'Times','Fontsize',12)
  end

    f7 = figure(7);
  for t =1:no_of_positions

  if(t<=8)
            y_v=0;
            plot(t,y_v,'b+','Linewidth',2)
  elseif(t>8)
            y_v=10;
            plot(t,y_v,'b+','Linewidth',2)
  end
        hold on
xlabel('Time, [s]','FontName','Times','Fontsize',12)
ylabel('Target Velocity in Y direction [m/sec]','FontName',
'Times','Fontsize',12)
  end
```

References

1. S.R. Jondhale, R.S. Deshpande, Kalman filtering framework based real time target tracking in wireless sensor networks using generalized regression neural networks. IEEE Sensors J. **19**, 224–233 (2018)
2. S. R. Jondhale, R. S. Deshpande, GRNN and KF framework based real time target tracking using PSOC BLE and smartphone. Ad Hoc Netw. (2019). https://doi.org/10.1016/j.adhoc.2018.09.017
3. S.R. Jondhale, R.S. Deshpande, Modified Kalman filtering framework based real time target tracking against environmental dynamicity in wireless sensor networks. Ad Hoc Sensor Wirel. Netw. **40**(1–2), 119–143 (2018)
4. T.K. Sarkar, Z. Ji, K. Kim, A. Medouri, M. Salazar-Palma, A survey of various propagation models for mobile communication. IEEE Antennas Propag. Mag. **45**, 51–82 (2003). https://doi.org/10.1109/MAP.2003.1232163
5. H. Wu, L. Zhang, Y. Miao, The propagation characteristics of radio frequency signals for wireless sensor networks in large-scale farmland. Wirel. Pers. Commun. (2017). https://doi.org/10.1007/s11277-017-4018-5
6. D. Balachander, T. R. Rao, G. Mahesh, RF propagation investigations in agricultural fields and gardens for wireless sensor communications, in *2013 IEEE Conference on Information & Communication Technologies* (2013). https://doi.org/10.1109/CICT.2013.6558195
7. A.S. Paul, E.A. Wan, RSSI-Based indoor localization and tracking using sigma-point kalman smoothers. IEEE J. Sel. Top. Signal Process. (2009). https://doi.org/10.1109/JSTSP.2009.2032309
8. S. Mahfouz, F. Mourad-Chehade, P. Honeine, J. Farah, H. Snoussi, Target tracking using machine learning and kalman filter in wireless sensor networks. IEEE Sensors J. (2014). https://doi.org/10.1109/JSEN.2014.2332098
9. X. R. Li, V. P. Jilkov, Survey of maneuvering target tracking. Part I: dynamic models. IEEE Trans. Aerosp. Electron. Syst. (2003). https://doi.org/10.1109/TAES.2003.1261132
10. F. Viani, P. Rocca, G. Oliveri, D. Trinchero, A. Massa, Localization, tracking, and imaging of targets in wireless sensor networks: an invited review. Radio Sci. (2011). https://doi.org/10.1029/2010RS004561

Chapter 6
GRNN-Based Target L&T Using RSSI

6.1 GRNN Architecture for Target L&T Applications

The ANNs not only have the capacity to handle noisy RSSI measurements, but they also have the ability to learn as well as get trained quickly [1–3]. In the applications wherein the system input-output mapping has a high amount of uncertainty due to environmental noise, the ANNs are widely used. As against KF framework, the application of the ANN does not necessitate prior knowledge of the noise distribution [4–6]. Hence, the ANN is one of the potential techniques in the target L&T. The ANN can very effectively deal with noisy RSSI measurements or the erroneous distances between wireless nodes. However, the choice of type of ANN for the given L&T application must be appropriate.

One-pass learning capability of the GRNN makes them able to deal with the underlying linear as well as nonlinear problem space very quickly [7, 8]. As compared to other ANNs, the GRNN can be trained quickly using very few training samples. The GRNN is able to compute how far the applied input vector from the training set patterns. The GRNN architecture is a type of supervised learning architecture. Its architecture consists of four types of layers, namely, input layer, summation layer, pattern layer, and output layer (see Fig. 6.1) [9–11]. Collecting the input signals is the responsibility of the nodes in the input layer, whereas pattern layer carries out the nonlinear transformation of the input data into the pattern space. In order to have a correct transformation, the count of the hidden layer neurons is generally kept equal to the count of the training set patterns. The summation layer carries out the summation, whereas the pattern layer node outputs are multiplied with the suitable interconnection weights to produce the network output.

The input to the proposed GRNN architecture is the RSSI measurements received from the four anchor nodes (i.e., *RSSI1, RSSI2, RSSI3, and RSSI4*), whereas the x and y coordinates of the mobile target are the output of the network (see Fig. 6.1).

S. R. Jondhale et al., *Received Signal Strength Based Target Localization and Tracking Using Wireless Sensor Networks*, EAI/Springer Innovations in Communication and Computing, https://doi.org/10.1007/978-3-030-74061-0_6

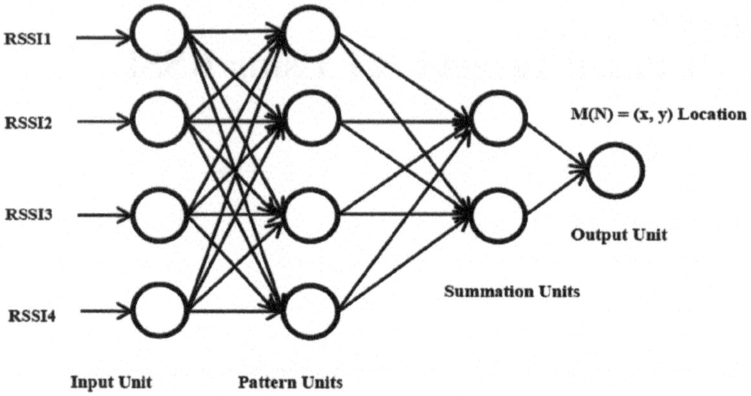

Fig. 6.1 GRNN architecture for target L&T

Technically speaking, the GRNN estimates the dependent variables M for any new independent variables N in a short amount of time as determined by the signal propagation time through it [4, 5]. In this chapter, M represents 2-D location to be estimated, whereas N is four RSSI measurements from four anchor nodes at a specific time instance. The mathematical relation between these two variables is given in Eq. (6.1):

$$M(N) = \frac{\sum_{i=1}^{m} M_i \exp\left(\frac{-D_i^2}{2\sigma^2}\right)}{\sum_{i=1}^{m} \exp\left(\frac{-D_i^2}{2\sigma^2}\right)} \tag{6.1}$$

The scalar function D_i^2 can be defined as the following:

$$D_i^2 = (N - N_i)^T . (N - N_i) \tag{6.2}$$

The variables σ and m used in Eq. (6.1) are known as spread factor of GRNN and dimension of input vector, respectively. Here in this chapter, $m = 4$. The choice of an appropriate value of m is very crucial from the performance accuracy point of view of the GRNN [9–12]. The higher the value of σ, the higher will be the ability of the underlying GRNN to generalize quickly and vice versa. Thus, the value of σ has to be chosen very carefully.

6.2 System Assumption and Design

The proposed target L&T system for this chapter is assumed to contain anchor nodes deployed at some fixed locations in the WSN of area 100 m by 100 m (see Figs. 6.4, 6.8, and 6.12). The base station is assumed to be located outside the WSN area. The mobile target carries a WSN node, which broadcasts RF signal to the surrounding anchor nodes at every time instance k. That means, it is assumed that the anchor nodes act as receivers and the target is a transmitter in the proposed experimental setup. It can be considered as a case of cooperative L&T. Each anchor node computes their distances from the mobile target using the collected RSSI measurements. The anchor nodes send these distances and their coordinates to the outside base station. The base station then selects the lowest three distances out of them. The base station is a laptop (processor: Core i3, CPU frequency 1.89 GHz, 2 GB RAM). Upon receiving the details from anchor nodes, the base station runs the trilateration algorithm to estimate the moving target location for each k time instance. The transmission power is assumed to be 1 mW, whereas the receiver and transmitter antenna gains are 1 dBm.

This chapter presents the investigation on three cases as described below:

- *Case I: Comparison of GRNN-based target L&T algorithm with traditional trilateration-based technique.*
- *Case II: Comparison of GRNN+KF and GRNN+UKF target L&T algorithms with GRNN- and trilateration-based L&T algorithms.*
- *Case III: Comparison of GRNN+KF and GRNN+UKF algorithms with trilateration+KF and trilateration+UKF L&T algorithms.*

In Case I, the target is assumed to go through the abrupt variations in the velocity during T sec as per Eqs. (6.3)–(6.9). This variation in velocity is depicted in Figs. 6.2 and 6.3:

$$\dot{x}_k = 1, \quad \dot{y}_k = 3, \qquad \text{for } 0 < k < 1 \text{ s}, \tag{6.3}$$

$$\dot{x}_k = 3, \quad \dot{y}_k = 7, \qquad \text{for } 1 \le k \le 9 \text{ s}, \tag{6.4}$$

$$\dot{x}_k = 2, \quad \dot{y}_k = 5, \qquad \text{for } 9 \le k \le 10 \text{ s}, \tag{6.5}$$

$$\dot{x}_k = 4, \quad \dot{y}_k = 2, \qquad \text{for } 10 \le k \le 15 \text{ s}, \tag{6.6}$$

$$\dot{x}_k = 0, \quad \dot{y}_k = 0, \qquad \text{for } 15 \le k \le 17 \text{ s}, \tag{6.7}$$

$$\dot{x}_k = 2, \quad \dot{y}_k = -3, \qquad \text{for } 17 \le k \le 29 \text{ s}, \tag{6.8}$$

$$\dot{x}_k = -2, \quad \dot{y}_k = -2, \qquad \text{for } 29 \le k \le 40 \text{ s}, \tag{6.9}$$

To incorporate this environmental dynamicity, the average values of n (n_{avg}) and $P_r(d_0)$ are computed in advance in the environmental calibration step (see Eqs. (6.10)–(6.12)). For the given three distances (d_1, d_2andd_3), and three RSSI measurements (z_3, z_2andz_3), the value of n (n_{avg}) is calculated as given below [4, 13]:

$$z_1 = P_r(d_0) - 10n_1 \log(d_1 / d_0) + X_\sigma, \tag{6.10}$$

$$z_2 = P_r(d_0) - 10n_2 \log(d_2 / d_0) + X_\sigma, \tag{6.11}$$

$$z_3 = P_r(d_0) - 10n_3 \log(d_3 / d_0) + X_\sigma. \tag{6.12}$$

where (n_1, n_2andn_3) are path loss exponents corresponding to distances d_1, $d_2$andd_3,respectively. By rearranging and subtracting the above equations, the values of n_1, $n_2$andn_3 can be easily computed. Thereafter, the average path loss exponent n_{avg} can be calculated by taking the average of n_1, $n_2$andn_3 as given below:

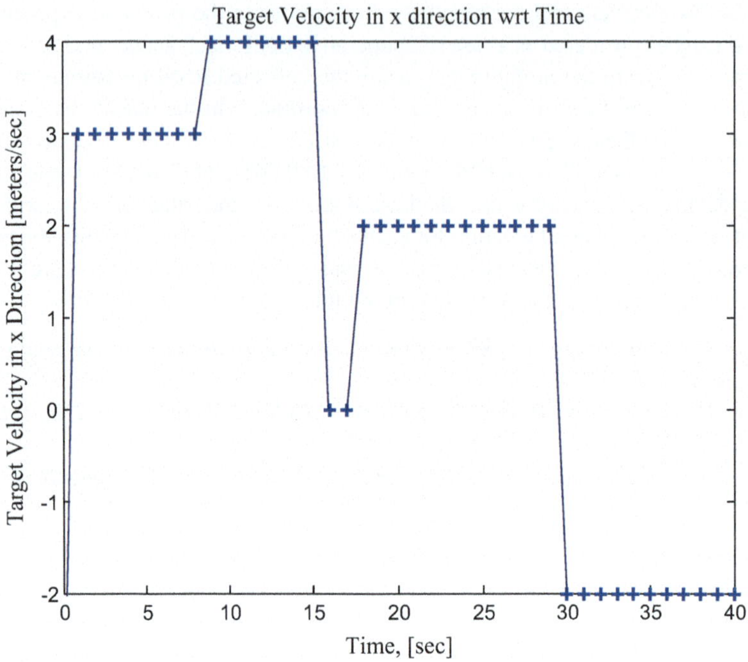

Fig. 6.2 Variation in target velocity along x direction

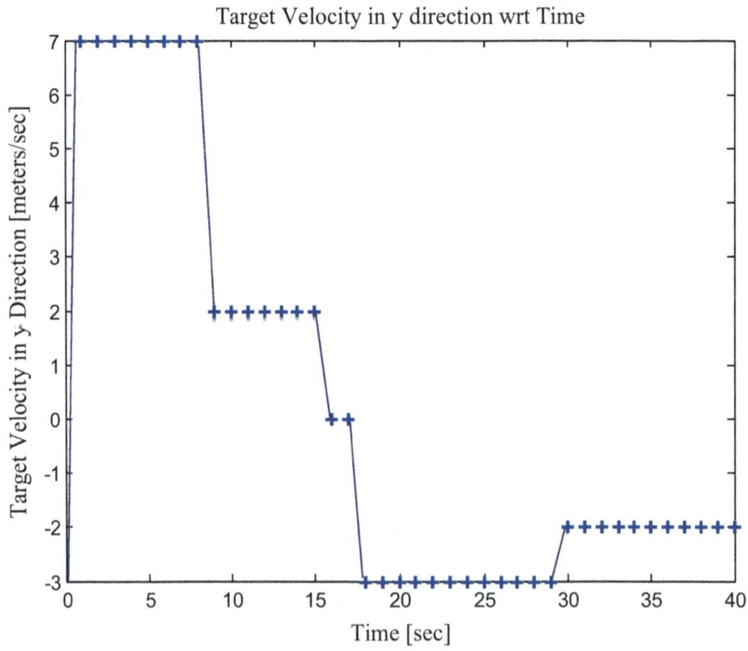

Fig. 6.3 Variation in target velocity along y direction

$$n_{avg} = (n_1 + n_2 + n_3)/3 \tag{6.13}$$

Therefore Eq. (6.10) is modified as

$$z_{lj,k} = P_r(d_0) - 10n_{avg}\log(d_{lj,k}/d_0) + X_\sigma . \tag{6.14}$$

The value of $P_r(d_0)$ and distance between receiver and transmitter can be determined using Eqs. (6.15) and (6.16), respectively:

$$P_r(d_0) = z_{lj,k} + 10n_{avg}\log(d_{lj,k}/d_0) - X_\sigma . \tag{6.15}$$

$$d_{lj,k} = d_0 10^{(P_r(d_0) - z_{lj,k} + X_\sigma)/10n_{avg}} . \tag{6.16}$$

The state of moving target at time instance k is given by the target state vector $X_k = (x_k, y_k, \dot{x}_k, \dot{y}_k)$, where x_k and y_k specify the position and $\dot{x}_k$ and $\dot{y}_k$ specify the velocity in x and y directions, respectively. All of the implementations in this chapter follow the CV mobility model. The target motion in all of these implementations is given by Eqs. (6.17) and (6.18):

$$x_k = x_{k-1} + \dot{x}\ dt , \tag{6.17}$$

$$y_k = y_{k-1} + \dot{y}\ dt . \tag{6.18}$$

The initial values of R, P and Q matrices are considered as follows:

$$R = \begin{bmatrix} 2.2 & 0 & 0 & 0 \\ 0 & 1.2 & 0 & 0 \\ 0 & 0 & 0.9 & 0 \\ 0 & 0 & 0 & 0.5 \end{bmatrix}, P = \begin{bmatrix} 0.25 & 0 & 0 & 0 \\ 0 & 0.4 & 0 & 0 \\ 0 & 0 & 0.2 & 0 \\ 0 & 0 & 0 & 0.01 \end{bmatrix}, Q = I_{4\times4} . \tag{6.19}$$

The initial target state vector for Case I, Case II, and Case III is assumed to be same. The detailed simulation parameters for these cases are as described in Table 6.1.

Table 6.1 Parameters assumed for simulation (Case I, Case II, and Case III)

Symbol	Parameter	Value
X_0	Target state vector at $k = 0$	[12 18 0 0]
–	Sensor communication range	100 m
dt	Discretization time step	1 s
–	Anchor density	4
T	Total simulation period	40 s
f	Frequency	2.4 GHz
X_σ	Normal random variable	~N (3, 1)
n_{avg}	Average path loss exponent	3.40

Table 6.2 Flow of GRNN+KF- and GRNN+UKF-based target L&T algorithms

I. Training of GRNN (off-line stage) *Step 1*: The proposed GRNN network is trained with 60 pairs of RSSI measurements from the anchor nodes and corresponding locations of target
II. Location estimation using GRNN (online stage) *Step 2*: The target receives RSSI values from anchor nodes for every *k*th time step, which are dispatched to the base station *Step 3*: The base station execute GRNN L&T algorithm to estimate target locations for every *k*th time step. The localization errors along *x* and *y* directions are recorded
III. Location estimation using KF (online stage) *Step 4*: The KF and UKF algorithms are used to refine estimates obtained in step 3 at the base station. The localization errors along x and y directionsare recorded For $k = 1, 2, \ldots T$ *Step 5*: Repeat step 2 to step4 for each next time instance *k*
IV. Determination of performance metrics *Step 6*: Calculate RMSE and average localization errorfor GRNN, trilateration, GRNN+KF and GRNN+UKF

6.3 Flow of Trilateration+KF- and Trilateration+UKF-Based L&T Algorithms

The algorithmic steps during time instance *k* in simulations contain three steps. The first step is environmental calibration (off-line step) (computation of n_{avg} and $P_r(d_0)$ empirically). The second part is the computation of distances, whereas the third part is the application of an input (smallest three distances between anchors and target (obtained in second part) and their coordinates) to trilateration algorithm. The obtained simulation results are based on the average of more than 50 trials of the proposed trilateration algorithm. The detailed flow of the trilateration-based L&T algorithm during *k* is as given in Table 6.2.

6.4 Performance Metrics

To assess the localization performance of trilateration, GRNN, trilateration+KF, trilateration+UKF, GRNN+KF, and GRNN+UKF algorithms, two performance metrics are used, namely. RMSE and average localization error [4, 13–16]. These metrics represent the average estimation error in target location $(\hat{x}_k, \hat{y}_k)$ and how much close the estimated target location $(\hat{x}_k, \hat{y}_k)$ and actual target location (x_k, y_k), respectively. These two metrics collectively represent a measure of target L&T accuracy. The lower the values of these parameters, the higher would be the target L&T accuracy. For each of the time instance *k*, the error in x estimate $(\hat{x}_k - x_k)$ and error in y estimate $(\hat{y}_k - y_k)$ for all of the above algorithms are computed for Case I and Case II. For each simulation run, the RMSE and average localization error are computed with the help of Eqs. (6.20) and (6.21), respectively [4–6, 17].

1. Average Localization Error

$$\text{Average Localization Error} = \frac{1}{T}\sum_{k=1}^{T}\frac{\left(\hat{x}_k - x_k\right)+\left(\hat{y}_k - y_k\right)}{2}, \tag{6.20}$$

2. RMSE

$$\text{RMSE} = \sqrt{\frac{1}{T}\sum_{k=1}^{T}\frac{\left(\hat{x}_k - x_k\right)^2+\left(\hat{y}_k - y_k\right)^2}{2}}. \tag{6.21}$$

6.5 Discussion on Results

6.5.1 Case I Results

Comparison of GRNN-based target L&T algorithm with traditional trilateration-based technique.

In Case I, the GRNN-based L&T algorithm is compared with traditional trilateration-based technique for the considered WSN setup. From Fig. 6.4, it is clear that the GRNN-based estimation of the target track is almost the same as that with the actual track of target as against trilateration. Figures 6.5, 6.6, and 6.7

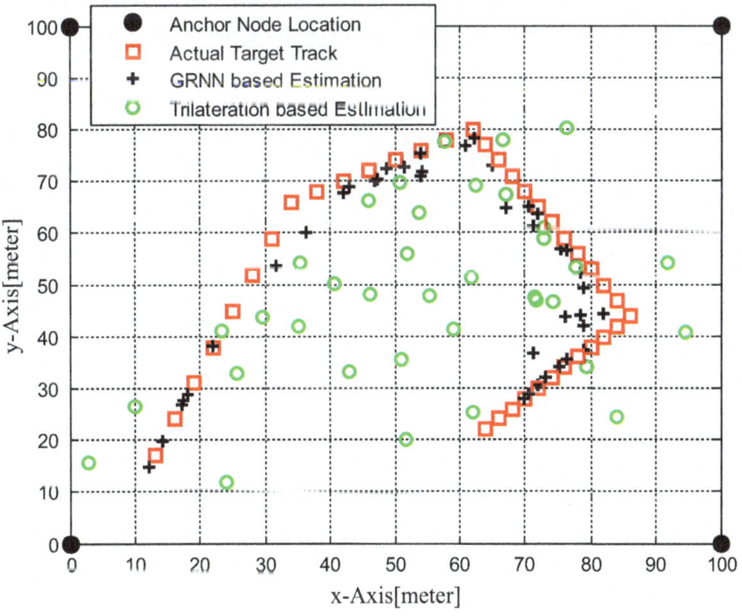

Fig. 6.4 Actual and estimated target trajectories with trilateration- and GRNN-based L&T algorithms (Case I)

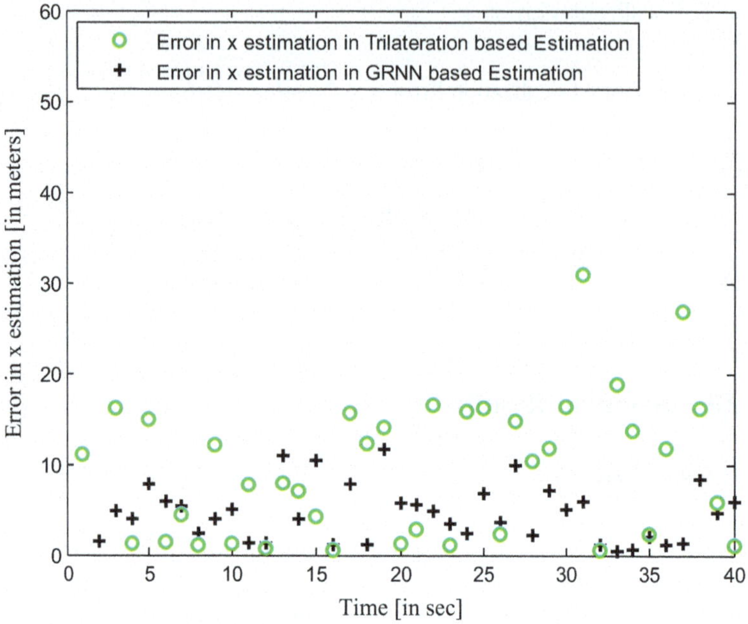

Fig. 6.5 Localization error in x estimate with trilateration and GRNN algorithms (Case I)

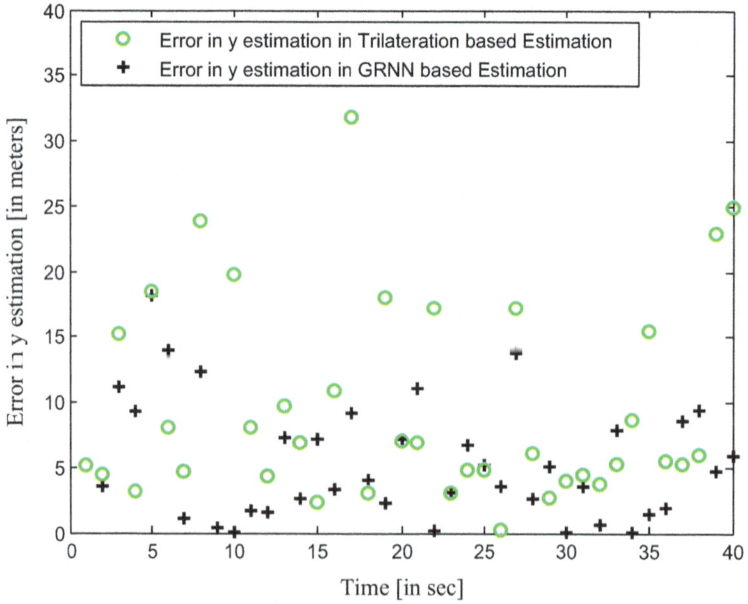

Fig. 6.6 Localization error in y estimate with trilateration and GRNN algorithms (Case I)

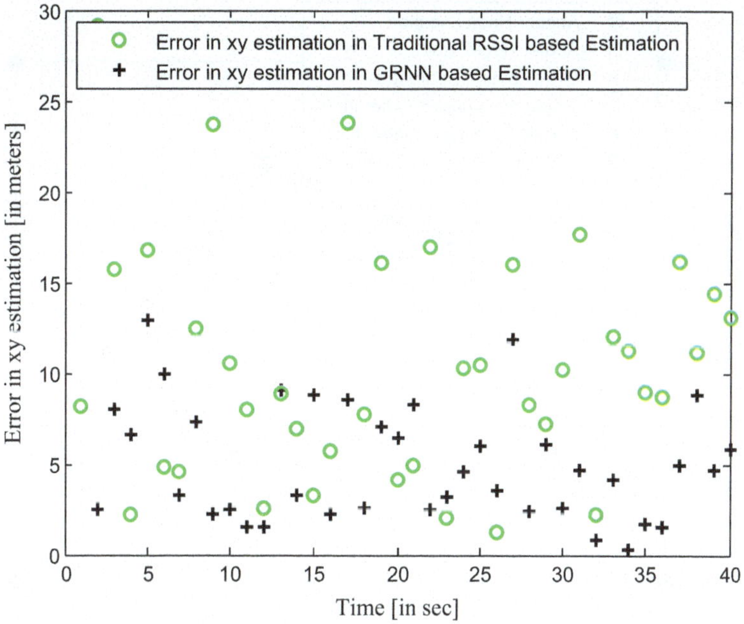

Fig. 6.7 Localization error in *x–y* estimate with trilateration and GRNN algorithms (Case I)

Table 6.3 Numeric results with trilateration- and GRNN-based L&T algorithms (Case I)

RSSI measurement noise = 3 dBm			
Sr. no.	Algorithm	Avg localization error	RMSE
1	Trilateration	10.4893	14.0275
2	GRNN	4.9188	6.1799

demonstrate the localization errors for each k values for both algorithms along *x*, *y*, and *x–y* directions, respectively. The localization errors with trilateration are connected by a solid red line in Figs. 6.5, 6.6, and 6.7.

With GRNN algorithm, the average localization error and the RMSE values are reduced by approximately 53% and 56% in comparison with trilateration (see Table 6.3). However, in some of L&T applications, there is a demand of high tracking accuracy than what is obtained in Case I (i.e., with GRNN alone). Hence, the GRNN-based location estimates are refined further using the popular KF framework (i.e., in Case II).

6.5.2 *Case II Results*

Comparison of GRNN+KF and GRNN+UKF target L&T algorithms with GRNN- and trilateration-based L&T algorithms.

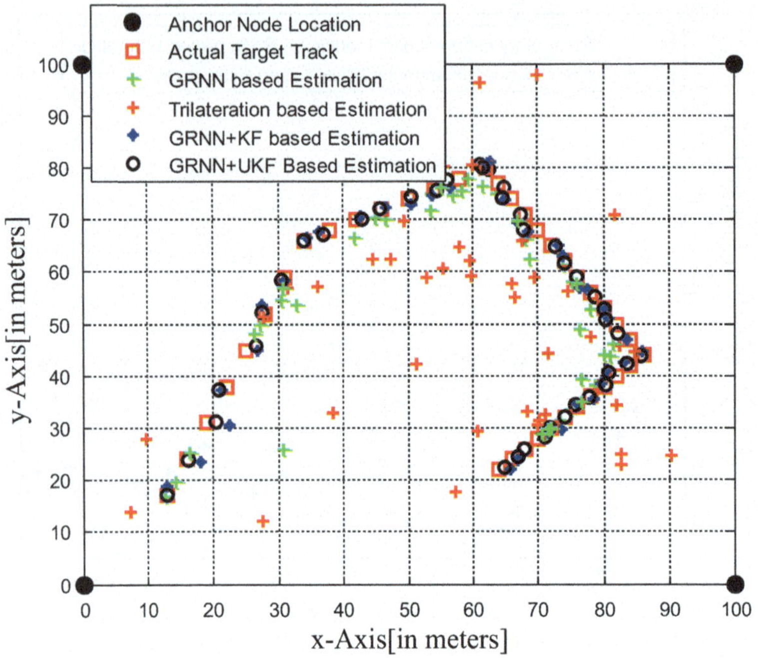

Fig. 6.8 Actual and estimated target tracks using trilateration-, GRNN-, GRNN+KF-, and GRNN+UKF-based L&T algorithms (Case II)

In Case II, the GRNN-based L&T algorithm is compared with trilateration, GRNN+KF, and GRNN+UKF algorithms for the same WSN setup. From Fig. 6.8, it is clear that the GRNN+UKF-based estimation of the target track is more accurate as compared to the rest of the implementations. Figures 6.9, 6.10, and 6.11 demonstrate the localization errors for each k values for all of the algorithms along *x*, *y*, and *x*–*y* directions, respectively. The simulation results prove that both the average localization error and RMSE values are below 1 m with GRNN+UKF as against the rest of the other implementations. These results are even promising as compared to Case I. The results obtained with GRNN+KF is also much better than GRNN- and trilateration-based estimations (see Table 6.3). With GRNN+UKF algorithm, the average localization error and the RMSE values are reduced by 94% and 93% in comparison with trilateration (see Table 6.4).

6.5.3 *Case III Results*

Comparison of GRNN+KF and GRNN+UKF target L&T algorithms with trilateration+KF and trilateration+UKF L&T algorithms.

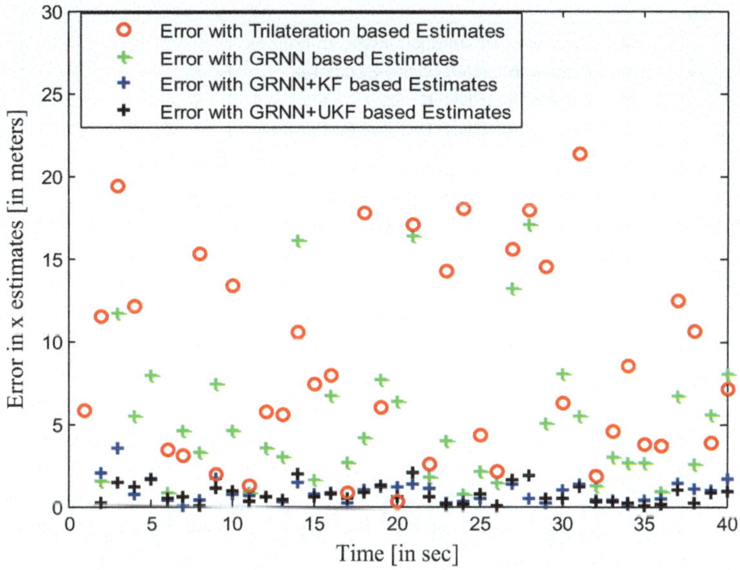

Fig. 6.9 Localization error in *x* estimate with trilateration-, GRNN-, GRNN+KF-, and GRNN+UKF-based L&T algorithms (Case II)

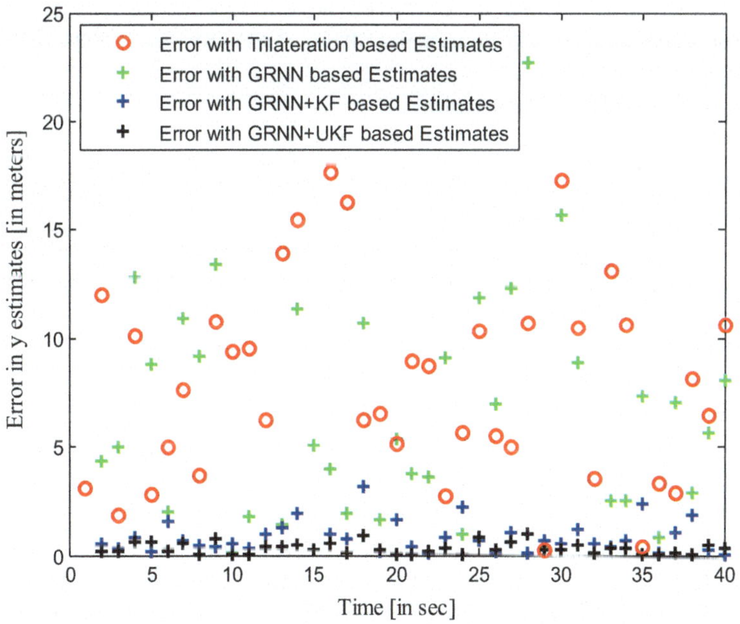

Fig. 6.10 Localization error in *y* estimate with trilateration-, GRNN-, GRNN+KF-, and GRNN+UKF-based L&T algorithms (Case II)

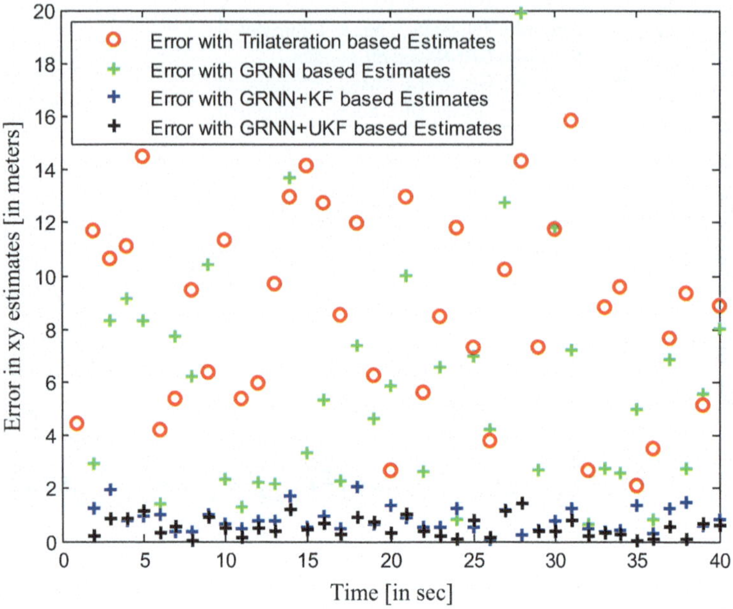

Fig. 6.11 Localization error in *x–y* estimate with trilateration-, GRNN-, GRNN+KF-, and GRNN+UKF-based L&T algorithms (Case II)

Table 6.4 Numeric results with trilateration-, GRNN-, GRNN+KF-, and GRNN+UKF-based L&T algorithms (Case II)

RSSI measurement noise = 3 dBm, sigma = 3.5			
Sr. no.	Algorithm	Avg localization error (in meters)	RMSE (in meters)
1	Trilateration	8.7027	10.4148
2	GRNN	5.6690	7.3640
3	GRNN+KF	0.8632	1.1133
4	GRNN+UKF	0.5160	0.6600

In Case III, the algorithms presented in Case II (i.e., GRNN+KF and GRNN+UKF) are compared with the previously defined trilateration+KF and trilateration+UKF algorithms again for the same considered WSN setup. From Fig. 6.12, it is clear that the GRNN+KF- as well as GRNN+UKF-based estimations perform much better than both trilateration+KF and trilateration+UKF algorithms in the context of target tracking. Figures 6.13, 6.14, and 6.15 demonstrate the localization errors for each k values for both algorithms along *x*, *y*, and *x–y* directions, respectively. The simulation results prove that both the average localization error and RMSE values for both GRNN+KF and GRNN+UKF are almost below 1 m with their trilateration-based counterparts (see Table 6.5).

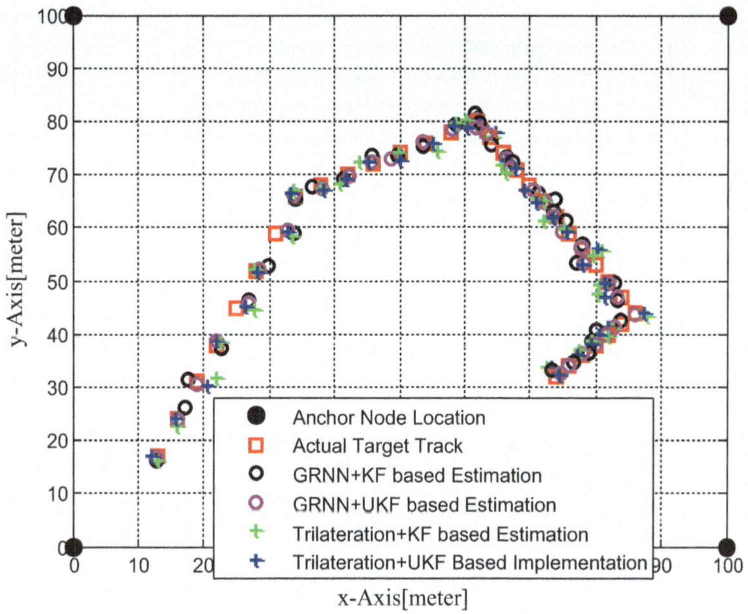

Fig. 6.12 Actual and estimated target tracks using GRNN+KF-, GRNN+UKF-, trilateration+KF-, and trilateration+UKF-based L&T algorithms (Case III)

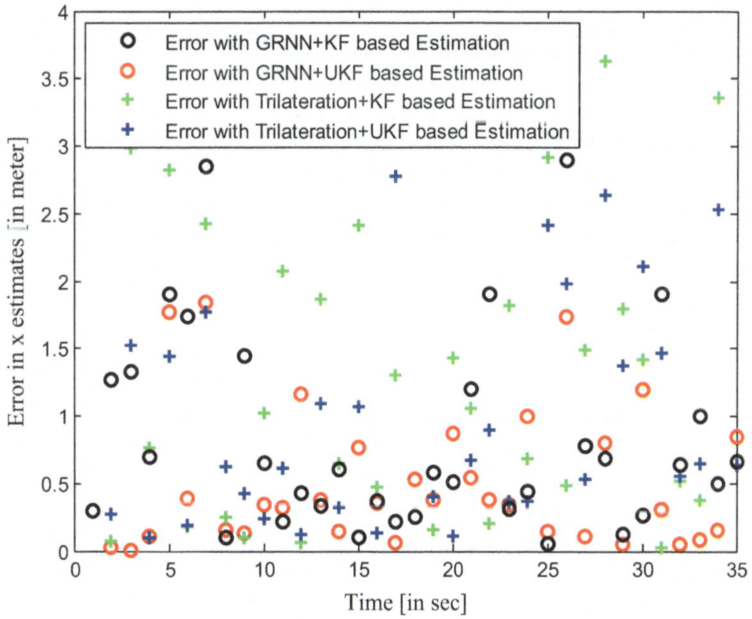

Fig. 6.13 Localization error in x estimate with GRNN+KF-, GRNN+UKF-, trilateration+KF-, and trilateration+UKF-based L&T algorithms (Case III)

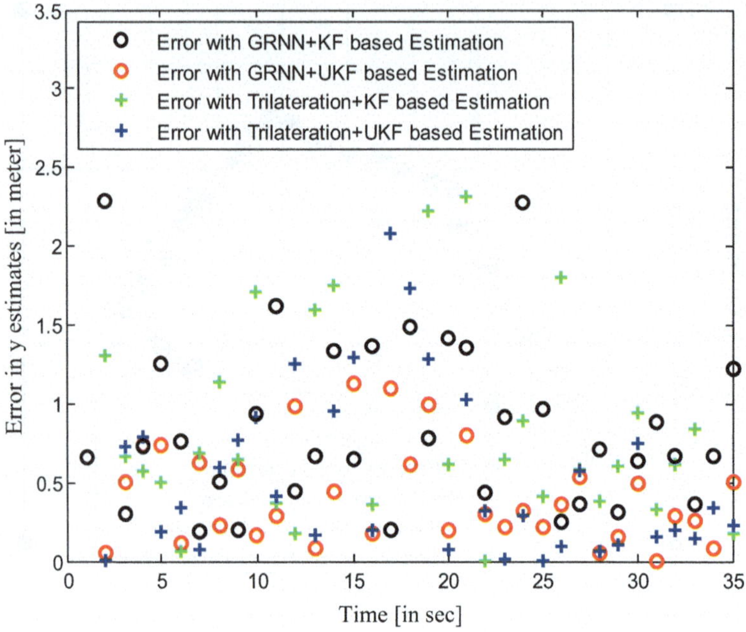

Fig. 6.14 Localization error in *y* estimate with GRNN+KF-, GRNN+UKF-, trilateration+KF-, and trilateration+UKF-based L&T algorithms (Case III)

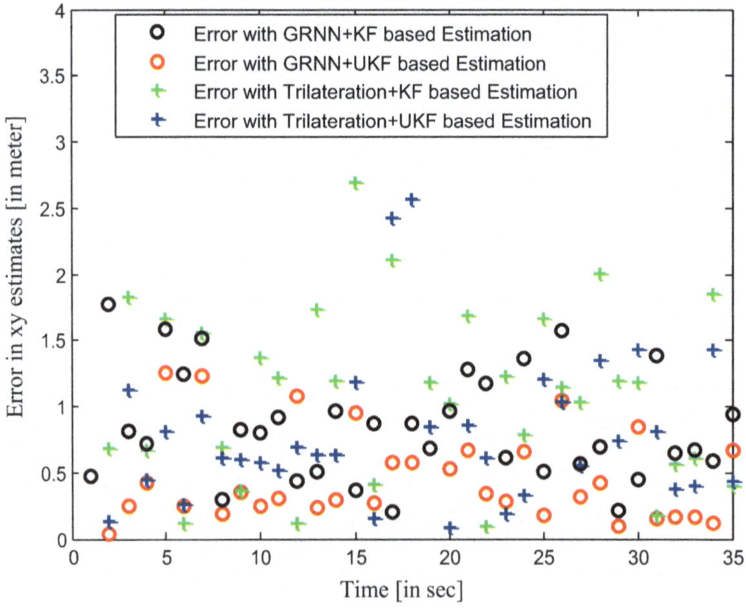

Fig. 6.15 Localization error in *x–y* estimate with GRNN+KF-, GRNN+UKF-, trilateration+KF-, and trilateration+UKF-based L&T algorithms (Case III)

Table 6.5 Numeric results with trilateration+KF-, trilateration+UKF-, GRNN+KF-, and GRNN+UKF-based L&T algorithms (Case III)

RSSI measurement noise = 3 dBm, sigma = 3.5			
Sr. no.	Algorithm	Avg localization error (in meters)	RMSE (in meters)
1	Trilateration+KF	1.1527	1.5235
2	Trilateration+KF	0.7849	1.0541
3	GRNN+KF	0.8440	1.0591
4	GRNN+UKF	0.4615	0.6184

6.6 Conclusions

The chapter discusses the application of the GRNN architecture coupled with KF framework for target L&T problem using WSN. Once trained with the RSSI measurements and associated locations in the off-line phase, it is able to learn the dynamicity of any given indoor environment quickly to give location estimates of the mobile target in the online phase. In this chapter, two algorithms GRNN+KF and GRNN+UKF are presented and are evaluated with extensive simulation experiments with the help of Case I, Case II, and Case III. In Case I, the performance of GRNN-based L&T algorithm is compared with the traditional trilateration-based localization technique. In Case II, the GRNN+KF and GRNN+UKF algorithms are compared with trilateration technique, whereas in Case III, the efficacy of GRNN+KF and GRNN+UKF algorithms are compared with the previously proposed trilateration+KF and trilateration+UKF algorithms. The proposed GRNN- and KF-based target L&T algorithms demonstrate superior target tracking performance irrespective of abrupt variations in the target velocity, environmental dynamicity, as well as nonlinear system dynamics. The research work claims the suitability of GRNN-based implementation over trilateration method.

MATLAB Codes for GRNN and KF Framework-Based Target L&T

MATLAB Code for Case I

%%%%%%%%%%%%% main.m %%%%%%%%%%%%%%%%%%%%%%%%%

```
%% WSN Deployment Setting Parameters
clearall
closeall
clc

networkSize = 100;
anchorLoc   = [0                0;          % set the anchor at 4 vertices of the region
networkSize         0;
                0          networkSize;
networkSizenetworkSize];

%%% GRNN Setting %%%%%%%%%%%%%%%%%%%%%%%%%%%%%%%%%%%%%%

RSSI_input_Vector = [-8.63711087700203,-11.5121738440855,-14.9788577741561,-
18.2424970326168,-17.2767505383621,-22.1529456887463,-20.5988132710408,-
26.7381951841692,-28.1016056626748,-26.8401681592955,-31.2009622756918,-
30.7067924858325,-27.0108637314813,-28.6009147429885,-34.8669185548126,-
33.6713973358156,-26.1945007053475,-34.1098185486167,-32.7378316422370,-
29.5555232963034,-29.1302982767054,-26.0425571563218,-31.0142194231226,-
23.9517855153775,-35.3809753208000,-30.8689220924318,-23.2519641062010,-
26.3382137588999,-33.2003851315704,-26.6366098384704,-26.8893597649793,-
28.1481934702383,-29.3271622723670,-29.3502853536700,-22.2650755453830,-
32.9478086658446,-19.1908979492906,-26.5517429750970,-27.8882690706260,-
24.0805034710059;-24.6443378867791,-23.8593396704728,-25.0300774444898,-
27.3552769837086,-31.3238874955404,-25.5684451777475,-27.2669962899477,-
28.5236948407618,-30.6095309413261,-27.3550681549064,-29.0951633851164,-
26.8496351417526,-25.2402279674739,-33.4366094367594,-28.2018648304488,-
31.7202588327514,-25.7273993271116,-27.7406442000865,-26.8123711146280,-
25.8920330826360,-32.8645336994140,-22.4362331219866,-22.7408893670739,-
26.6055760301610,-15.1061425078805,-22.8393577737386,-16.5018015844959,-
20.3573928149789,-25.9018275788868,-19.0747887851302,-16.4556233464885,-
17.9921345751942,-18.2818096318861,-18.0160361661470,-21.0946978981921,-
14.8199088088597,-17.4389955511778,-14.5491119152686,-16.9372270707295,-
21.7719689162081;-28.5135613143206,-33.0780989357883,-31.3506779461191,-
36.0066734173278,-36.4771716278969,-26.2060894708195,-31.9861022431075,-
25.0584109759550,-28.0201092305810,-20.7125177645295,-20.6950504734078,-
20.6937065669968,-25.6230277199468,-18.7990725652173,-19.3405715045320,-
13.7283029756786,-15.7661646303376,-16.9170765891923,-12.4096156327949,-
20.4488014567320,-11.2793351657313,-19.2735444611160,-12.9127466593933,-
```

```
16.6973307478935,-23.2440861614554,-19.5423962483160,-17.8774561543769,-
22.2394269412980,-23.2321402317115,-18.2863155370521,-22.0255637228001,-
25.5137093113793,-26.9308086311955,-21.8929688622825,-19.7266313616989,-
32.6206155831883,-24.9869109502726,-28.0621473668449,-28.2854211807112,-
32.7967125887547;-25.9118064691782,-27.2885447453488,-23.5263426077213,-
19.7371525265260,-17.6956931946208,-20.1876598879298,-18.2350285824841,-
18.2568318614458,-16.4786757425207,-23.8035181257022,-23.5704645802868,-
19.9652392972093,-21.9046157843161,-19.8548782223887,-21.6613529489605,-
22.3355342877953,-23.4218741164849,-22.8986277147529,-26.6970793451071,-
24.5674064740980,-22.1249165070594,-30.4306177007705,-30.4071198812238,-
28.1721577460586,-30.6151353968228,-28.0296173820039,-31.5003675294334,-
28.3877986063688,-33.0397705487563,-32.0049067239378,-33.2699125292289,-
31.2200071293077,-33.9181605899042,-31.9872995878781,-29.6661170371290,-
29.1643624501848,-28.1690529175130,-27.5642192144777,-32.5581354439923,-
34.6179330646867];
Target =
[10,13,16,19,22,25,28,31,34,38,42,46,50,54,58,62,62,62,64,66,68,70,72,74,76,78,80,82,84,86,84
,82,80,78,76,74,72,70,68,66;10,17,24,31,38,45,52,59,66,68,70,72,74,76,78,80,80,80,77,74,71,68
,65,62,59,56,53,50,47,44,42,40,38,36,34,32,30,28,26,24];
spread =3.5;

net_Loc_est = newgrnn(RSSI_input_Vector,Target,spread);
view(net_Loc_est)


%show anchor Locations
   f1 = figure(1);
   plot(anchorLoc(:,1),anchorLoc(:,2),'ko','MarkerSize',8,'lineWidth',2,'MarkerFaceColor','k');
axis([0 100 0 100])
gridon
holdon

% Definingveriables
no_of_positions = 40;     % Total Simulation Period = 35 seconds

%%%%%%%%%%%%%%% Initialization of Error Values %%%%%%%%%%%%%%%%%
RMSE_rssi_x= 0; RMSE_grnn_x = 0;
RMSE_rssi_y= 0; RMSE_grnn_y = 0;

% Calculate reference RSSI at d0 = 1 meter using Free Space Path Loss Model
   d0=1;                          % in Meters
   Pr0 = RSSI_friss(d0);

d_test = 20;

Pr = RSSI_friss(d_test);
```

```
%Calculation of Path Loss Exponent :

    n = -(Pr + Pr0)/(10*log(d_test));
    x=10;
    y=10;

% Generating trajectory for the mobile node
for t = 1:no_of_positions
if(t<9)                    x_v = 3; y_v = 7;
elseif(t==9 && t<16)    x_v = 4; y_v = 2;
elseif(t>=16 && t<18)  x_v = 0; y_v = 0;
elseif(t>=18 && t<30)  x_v = 2; y_v = -3;
elseif(t>=30 && t<=40) x_v = -2; y_v = -2;
end

x_actual(t)=x;       % para 1
y_actual(t)=y;       % para 1

     x=x+x_v;
     y=y+y_v;

disp('True Location')
     [x,y]                          % Actual Target Location
plot(x,y,'rs','LineWidth',2)

ylabel('y-Axis[meter]','FontName','Times','Fontsize',14,'LineWidth',2);
xlabel('x-Axis[meter]','FontName','Times','Fontsize',14,'LineWidth',2);
legend('Anchor Node Location','Actual Target Track','GRNN based Estimation',
'Trilateration based Estimation','Location','SouthEast')
holdon

% Actual Distances from Anchors required to generate RSSI Values
     d1 = sqrt( x^2 + y^2 );
     d2 = sqrt((100-x)^2 + y^2);
     d3 = sqrt((100-x)^2+ (100-y)^2);
     d4 = sqrt(x^2+ (100-y)^2);

% Generate RSSI Values at 4 Anchor Nodes which are at d1, d2, d3 & d4 distances respectively
from Moving Target

     RSS = lognormalshadowing_4(n,d1,d2,d3,d4,Pr0);

     RSS_1(t)= RSS(1);
     RSS_2(t)= RSS(2);
     RSS_3(t)= RSS(3);                %4 RSS Values
     RSS_4(t)= RSS(4);


RSS_new_vector = RSS.';
GRNN_Estimated_Loc = sim(net_Loc_est ,RSS_new_vector)
```

```
GRNN_x(t)= GRNN_Estimated_Loc(1);
GRNN_y(t)= GRNN_Estimated_Loc(2);
plot(GRNN_Estimated_Loc(1),GRNN_Estimated_Loc(2),'k+','LineWidth',2)
holdon

    RSS_s = sort(RSS);

disp('RSSI Estimated Location')

mobileLoc_est = trilateration_4(RSS,RSS_s,Pr0,n,networkSize)

    X_T = mobileLoc_est(1);
    Y_T = mobileLoc_est(2);

trad_x(t)=X_T;
trad_y(t)=Y_T;

plot(X_T,Y_T,'go','LineWidth',2)
holdon


% Error Analysis of algorithm
% ---> Part 1 : RMSE Analysis
RMSE_rssi_x = RMSE_rssi_x + (X_T - x)^2 ;
RMSE_rssi_y = RMSE_rssi_y+  (Y_T - y)^2;
RMSE_grnn_x = RMSE_grnn_x + (GRNN_Estimated_Loc(1) - x)^2 ;
RMSE_grnn_y = RMSE_grnn_y + (GRNN_Estimated_Loc(2) - y)^2;

% ---> Part 2 : Calculation of Absolute Errors

% a) For Trilateration based Technique
error_x_rssi(t) = abs((x - X_T));
error_y_rssi(t) = abs((y - Y_T));
error_xy_rssi(t) =((error_x_rssi(t) + error_y_rssi(t))/2);


% b) For GRNN based Estimation
error_x_grnn(t) = abs((x - GRNN_Estimated_Loc(1)));
error_y_grnn(t) = abs((y - GRNN_Estimated_Loc(2)));
error_xy_grnn(t) =((error_x_grnn(t) + error_y_grnn(t))/2);
end

% Average Error in x &y  coordinates
avg_error_xy_rssi = 0;
avg_error_xy_grnn = 0;

for t = 1:no_of_positions
avg_error_xy_rssi=avg_error_xy_rssi + (error_xy_rssi(t)/no_of_positions);
avg_error_xy_grnn=avg_error_xy_grnn + (error_xy_grnn(t)/no_of_positions);
end
```

```
disp('Average Localization Errors :')
avg_error_xy_rssi
avg_error_xy_grnn

disp('RMSE Errors :')
RMSE_rssi_x = sqrt(RMSE_rssi_x/no_of_positions)
RMSE_rssi_y = sqrt(RMSE_rssi_y/no_of_positions)
RMSE_grnn_x = sqrt(RMSE_grnn_x/no_of_positions)
RMSE_grnn_y = sqrt(RMSE_grnn_y/no_of_positions)

RMSE_rssi_avg = (RMSE_rssi_x + RMSE_rssi_y)/2
RMSE_grnn_avg = (RMSE_grnn_x + RMSE_grnn_y)/2

% Plotting Absolute Errors of KF & UKF based Tracking

f2 = figure(2);
for t =1:no_of_positions

plot(t,error_x_grnn(t),'k+','LineWidth',2)
plot(t,error_x_rssi(t),'go','LineWidth',2)

xlabel('Time [in sec]','FontName','Times','Fontsize',14,'LineWidth',2)
ylabel('Error in x estimation [in meters]','FontName','Times','Fontsize',14,'LineWidth',2)
holdon

end
legend('Error in x estimation in Trilateration based Estimation','Error in x estimation in GRNN
based Estimation','Location','NorthWest')

  f3 = figure(3);
for t =1:no_of_positions

plot(t,error_y_grnn(t),'k+','LineWidth',2)
plot(t,error_y_rssi(t),'go','LineWidth',2)   %,'Markersize',2,'MarkerEdgeColor','g')

xlabel('Time [in sec]','FontName','Times','Fontsize',14, 'LineWidth',2)
ylabel('Error in y estimation [in meters]','FontName','Times','Fontsize',14, 'LineWidth',2)

holdon
end
legend('Error in y estimation in Trilateration based Estimation','Error in y estimation in GRNN
based Estimation','Location','NorthWest')

  f4 = figure(4);
for t =1:no_of_positions

plot(t,error_xy_grnn(t),'k+','LineWidth',2)
plot(t,error_xy_rssi(t),'go','LineWidth',2)   %,'Markersize',2,'MarkerEdgeColor','g')
```

```
xlabel('Time [in sec]','FontName','Times','Fontsize',14,'LineWidth',2)
ylabel('Error in xy estimation [in meters]','FontName','Times','Fontsize',14,'LineWidth',2)

holdon
end
legend('Error in xy estimation in Traditional RSSI based Estimation','Error in xy estimation in
GRNN based Estimation','Location','NorthWest')


 f5 = figure(5);
for t =1:no_of_positions

if(t<9)              x_v = 3; y_v = 7;
elseif(t==9 && t<16)   x_v = 4; y_v = 2;
elseif(t>=16 && t<18)  x_v = 0; y_v = 0;
elseif(t>=18 && t<30)  x_v = 2; y_v = -3;
elseif(t>=30 && t<=40) x_v = -2; y_v = -2;
end

plot(t,x_v,'b+','Linewidth',2)     %,'Markersize',2,'MarkerEdgeColor','b')
xlabel('Time, [sec]','FontName','Times','Fontsize',14)
ylabel('Target Velocity in x Direction [meters/sec]','FontName','Times','Fontsize',14)
title('Target Velocity in x direction wrt Time','FontName','Times','FontSize',12);
holdon
end

 f6 = figure(6);
for t =1:no_of_positions

if(t<9)              x_v = 3; y_v = 7;
elseif(t==9 && t<16)   x_v = 4; y_v = 2;
elseif(t>=16 && t<18)  x_v = 0; y_v = 0;
elseif(t>=18 && t<30)  x_v = 2; y_v = -3;
elseif(t>=30 && t<=40) x_v = -2; y_v = -2;
end

plot(t,y_v,'b+','Linewidth',2)     %,'Markersize',2,'MarkerEdgeColor','b')
xlabel('Time [sec]','FontName','Times','Fontsize',14)
ylabel('Target Velocity in y Direction [meters/sec]','FontName','Times','Fontsize',14)
title('Target Velocity in y direction wrt Time','FontName','Times','FontSize',12);
holdon
end
```

%%%%%%%%%%%%%%%%%%%%%% RSSI_friss.m %%%%%%%%%%%%

```
function [ Pr ] = RSSI_friss( d )

C=3e8;                  %LightSpeed
prompt = 'Enter Frequency in MHz : ';
Freq = input(prompt);

Freq = Freq*1000000;
%Freq=2400*1000000;%hz
%Freq = 867*1000000;
Zigbee=915.0e6;%hz
TXAntennaGain=1;%db
RXAntennaGain=1;%db

PTx=0.001;%watt

%%%%%%  FRIIS Equation %%%%%%%%%

%      Pt * Gt * Gr * (Wavelength^2)
%  Pr = --------------------------
%      (4 *pi * d)^2 * L

    Wavelength=C/Freq;
PTxdBm=10*log10(PTx*1000);
    M = Wavelength / (4 * pi * d);
Pr=PTxdBm + TXAntennaGain + RXAntennaGain- (20*log10(1/M));    % Pr0 means A in RSSI
formula

end
```

%%%%%%%%%%%%%% lognormalshodowing_4.m %%%%%%%%%%%%

```
function [ RSS ] = lognormalshadowing_4( n,d1,d2,d3,d4,Pr0 )
%UNTITLED5 Summary of this function goes here
%  Detailed explanation goes here
%clear all
std_meas_noise = 3;  % standard deviation 0f measurement noise in db


    RSS_d1= -(10*n* log10(d1) + Pr0) + std_meas_noise*randn;
    RSS_d2= -(10*n* log10(d2) + Pr0) + std_meas_noise*randn;
    RSS_d3= -(10*n* log10(d3) + Pr0) + std_meas_noise*randn;
    RSS_d4= -(10*n* log10(d4) + Pr0) + std_meas_noise*randn;

    RSS = [ RSS_d1, RSS_d2, RSS_d3, RSS_d4];
end
```

%%%%%%%%%%%%%% trilateration_4.m %%%%%%%%%%%%%%

```
function [ mobileLoc_est ] = trilateration_4( RSS,RSS_s,Pr0,n,networkSize )

        d1_est = 10^(-(Pr0+RSS_s(4))/(10*n));
        d2_est = 10^(-(Pr0+RSS_s(3))/(10*n));
        d3_est = 10^(-(Pr0+RSS_s(2))/(10*n));

if (RSS_s(4) == RSS(1))
        X1 = 0; Y1 = 0;
elseif (RSS_s(4) == RSS(2))
        X1 = networkSize; Y1 = 0;
elseif (RSS_s(4) == RSS(3))
        X1 = networkSize; Y1 = networkSize;
elseif (RSS_s(4) == RSS(4))
        X1 = 0; Y1 = networkSize;
end

if (RSS_s(3) == RSS(1))
        X2 = 0; Y2 = 0;
elseif (RSS_s(3) == RSS(2))
        X2 = networkSize; Y2 = 0;
elseif (RSS_s(3) == RSS(3))
        X2 = networkSize; Y2 = networkSize;
elseif (RSS_s(3) == RSS(4))
        X2 = 0; Y2 = networkSize;
end
```

```
if (RSS_s(2) == RSS(1))
        X3 = 0; Y3 = 0;
elseif (RSS_s(2) == RSS(2))
        X3 = networkSize; Y3 = 0;
elseif (RSS_s(2) == RSS(3))
        X3 = networkSize; Y3 = networkSize;
elseif (RSS_s(2) == RSS(4))
        X3 = 0; Y3 = networkSize;
end


    A = X1^2 + Y1^2 - d1_est^2;
    B = X2^2 + Y2^2 - d2_est^2;
    C = X3^2 + Y3^2 - d3_est^2;

    X32 = (X3 - X2);
    Y32 = (Y3 - Y2);
    X21 = (X2 - X1);
    Y21 = (Y2 - Y1);
    X13 = (X1 - X3);
    Y13 = (Y1 - Y3);

    X_T = (A*Y32 + B*Y13 + C*Y21) / (2*( X1*Y32 + X2*Y13 + X3*Y21));
% X coordinate of target

    Y_T = (A*X32 + B*X13 + C*X21) / (2*( Y1*X32 + Y2*X13 + Y3*X21));
% Y coordinate of target

mobileLoc_est = [ X_T, Y_T];

end
```

MATLAB Code for Case II

```
%% WSN Deployment Setting Parameters
clearall
closeall
clc

   N = 4;                              % number of anchors
   M = 1;                              % number of mobile nodes
networkSize = 100;
anchorLoc  = [0              0;         % set the anchor at 4 vertices of the region
networkSize        0;
                0        networkSize;
networkSizenetworkSize];

%%% GRNN Setting %%%%%%%%%%%%%%%%%%%%%%%%%%%%%%%%%%%%%%%%%%%%%%%%%%%%%%

RSSI_input_Vector = [-8.63711087700203,-11.5121738440855,-14.9788577741561,-
18.2424970326168,-17.2767505383621,-22.1529456887463,-20.5988132710408,-
26.7381951841692,-28.1016056626748,-26.8401681592955,-31.2009622756918,-
30.7067924858325,-27.0108637314813,-28.6009147429885,-34.8669185548126,-
33.6713973358156,-26.1945007053475,-34.1098185486167,-32.7378316422370,-
29.5555232963034,-29.1302982767054,-26.0425571563218,-31.0142194231226,-
23.9517855153775,-35.3809753208000,-30.8689220924318,-23.2519641062010,-
26.3382137588999,-33.2003851315704,-26.6366098384704,-26.8893597649793,-
28.1481934702383,-29.3271622723670,-29.3502853536700,-22.2650755453830,-
32.9478086658446,-19.1908979492906,-26.5517429750970,-27.8882690706260,-
24.0805034710059;-24.6443378867791,-23.8593396704728,-25.0300774444898,-
27.3552769837086,-31.3238874955404,-25.5684451777475,-27.2669962899477,-
28.5236918407618,-30.6095309413261,-27.3550681549064,-29.0951633851164,-
26.8496351417526,-25.2402279674739,-33.4366094367594,-28.2018648304488,-
31.7202588327514,-25.7273993271116,-27.7406442000865,-26.8123711146280,-
25.8920330826360,-32.8645336994140,-22.4362331219866,-22.7408893670739,-
26.6055760301610,-15.1061425078805,-22.8393577737386,-16.5018015844959,-
20.3573928149789,-25.9018275788868,-19.0747887851302,-16.4556233464885,-
17.9921345751942,-18.2818096318861,-18.0160361661470,-21.0946978981921,-
14.8199088088597,-17.4389955511778,-14.5491119152686,-16.9372270707295,-
21.7719689162081;-28.5135613143206,-33.0780989357883,-31.3506779461191,-
36.0066734173278,-36.4771716278969,-26.2060894708195,-31.9861022431075,-
25.0584109759550,-28.0201092305810,-20.7125177645295,-20.6950504734078,-
20.6937065669968,-25.6230277199468,-18.7990725652173,-19.3405715045320,-
13.7283029756786,-15.7661646303376,-16.9170765891923,-12.4096156327949,-
20.4488014567320,-11.2793351657313,-19.2735444611160,-12.9127466593933,-
16.6973307478935,-23.2440861614554,-19.5423962483160,-17.8774561543769,-
22.2394269412980,-23.2321402317115,-18.2863155370521,-22.0255637228001,-
25.5137093113793,-26.9308086311955,-21.8929688622825,-19.7266313616989,-
32.6206155831883,-24.9869109502726,-28.0621473668449,-28.2854211807112,-
```

```
32.7967125887547;-25.9118064691782,-27.2885447453488,-23.5263426077213,-
19.7371525265260,-17.6956931946208,-20.1876598879298,-18.2350285824841,-
18.2568318614458,-16.4786757425207,-23.8035181257022,-23.5704645802868,-
19.9652392972093,-21.9046157843161,-19.8548782223887,-21.6613529489605,-
22.3355342877953,-23.4218741164849,-22.8986277147529,-26.6970793451071,-
24.5674064740980,-22.1249165070594,-30.4306177007705,-30.4071198812238,-
28.1721577460586,-30.6151353968228,-28.0296173820039,-31.5003675294334,-
28.3877986063688,-33.0397705487563,-32.0049067239378,-33.2699125292289,-
31.2200071293077,-33.9181605899042,-31.9872995878781,-29.6661170371290,-
29.1643624501848,-28.1690529175130,-27.5642192144777,-32.5581354439923,-
34.6179330646867];

Target =
[10,13,16,19,22,25,28,31,34,38,42,46,50,54,58,62,62,62,64,66,68,70,72,74,76,78,80,82,84,86,84
,82,80,78,76,74,72,70,68,66;10,17,24,31,38,45,52,59,66,68,70,72,74,76,78,80,80,80,77,74,71,68
,65,62,59,56,53,50,47,44,42,40,38,36,34,32,30,28,26,24];
spread =3.5;

net_Loc_est = newgrnn(RSSI_input_Vector,Target,spread);
view(net_Loc_est)


%show anchor Locations
   f1 = figure(1);
   plot(anchorLoc(:,1),anchorLoc(:,2),'ko','MarkerSize',8,'lineWidth',2,'MarkerFaceColor','k');
axis([0 100 0 100])
gridon
holdon

% Definingveriables
no_of_positions = 40;    % Total Simulation Period = 35 seconds


   P = [0.25 0 0  0; 0 0.04 0 0; 0 0 0.02 0; 0 0 0 0.01];  % Initial Process Covariance Matrix
   q=0.1;                                    %std of process

%%%%%%%%%%%%% Initialization of Error Values %%%%%%%%%%%%%%%%
RMSE_kf_x =0 ;RMSE_rssi_x= 0; RMSE_ukf_x= 0; RMSE_grnn_x = 0;
RMSE_kf_y =0 ;RMSE_rssi_y= 0; RMSE_ukf_y= 0; RMSE_grnn_y = 0;


% Calculate reference RSSI at d0 = 1 meter using Free Space Path Loss Model
   d0=1;                                     % in Meters
   Pr0 = RSSI_friss(d0);

d_test = 20;
Pr = RSSI_friss(d_test);

%Calculation of Path Loss Exponent :

    n = -(Pr + Pr0)/(10*log(d_test));
    x=10;
    y=10;
```

```
% Generating trajectory for the mobile node
for t = 1:no_of_positions
if(t<9)             x_v = 3; y_v = 7;
elseif(t==9 && t<16)   x_v = 4; y_v = 2;
elseif(t>=16 && t<18)  x_v = 0; y_v = 0;
elseif(t>=18 && t<30)  x_v = 2; y_v = -3;
elseif(t>=30 && t<=40)  x_v = -2; y_v = -2;
end

x_actual(t)=x;      % para 1
y_actual(t)=y;      % para 1

    x=x+x_v;
    y=y+y_v;

disp('True Location')
    [x,y]                           % Actual Target Location
plot(x,y,'rs','LineWidth',2)

ylabel('y-Axis[in meters]','FontName','Times','Fontsize',14,'LineWidth',2);
xlabel('x-Axis[in meters]','FontName','Times','Fontsize',14,'LineWidth',2);
    legend('Anchor Node Location','Actual Target Track','GRNN based
Estimation','Trilateration based Estimation','GRNN+KF based Estimation','GRNN+UKF Based
Estimation','Location','SouthEast')
holdon

% Actual Distances from Anchors required to generate RSSI Values
    d1 = sqrt( x^2 + y^2 );
    d2 = sqrt((100-x)^2 + y^2);
    d3 = sqrt((100-x)^2+ (100-y)^2);
    d4 = sqrt(x^2+ (100-y)^2);

% Generate RSSI Values at 4 Anchor Nodes which are at d1, d2, d3 & d4 distances respectively
from Moving Target
% Use  RSSI = - (10*n*log10(d) + A)  and  d= antilog(-(RSSI + A)/(10*n))

    RSS = lognormalshadowing_4(n,d1,d2,d3,d4,Pr0);
    RSS_1(t)= RSS(1);
    RSS_2(t)= RSS(2);
    RSS_3(t)= RSS(3);                  %4 RSS Values
    RSS_4(t)= RSS(4);

RSS_new_vector = RSS.';
GRNN_Estimated_Loc = sim(net_Loc_est ,RSS_new_vector)

GRNN_x(t)= GRNN_Estimated_Loc(1);
GRNN_y(t)= GRNN_Estimated_Loc(2);
plot(GRNN_Estimated_Loc(1),GRNN_Estimated_Loc(2),'g+','LineWidth',2)
holdon

    RSS_s = sort(RSS);

disp('RSSI Estimated Location')
```

```
mobileLoc_est = trilateration_4(RSS,RSS_s,Pr0,n,networkSize)

      X_T = mobileLoc_est(1);
      Y_T = mobileLoc_est(2);

trad_x(t)=X_T;
trad_y(t)=Y_T;

plot(X_T,Y_T,'r+','LineWidth',2)
holdon

% calculate velocities in X & Y Directions

velocity_est=velocity(X_T,Y_T,t);

        X = [x; y; 0; 0];            % State of the system at time 't'


% Kalman Filter for Tracking Moving Target Code starts here
      [X_kalman,Y_kalman,X_kf]= kf(X,P,GRNN_Estimated_Loc(1), GRNN_Estimated_Loc(2),
velocity_est,t);

kf_x(t)=X_kalman;    %Para 3
kf_y(t)=Y_kalman;    %Para 4

plot(X_kalman,Y_kalman,'b*','LineWidth',2)

holdon

%UKF implementation
     Z = [GRNN_Estimated_Loc(1); GRNN_Estimated_Loc(2); velocity_est(1); velocity_est(2)];
disp('UKF Estimated State')
     [X_ukf]= ukf5(X,P,Z)

ukf_x(t)=X_ukf(1);     %Para 5
ukf_y(t)=X_ukf(2);     %Para 6

plot(X_ukf(1),X_ukf(2),'ko','LineWidth',2)

holdon

% Error Analysis of algorithm
% ---> Part 1 : RMSE Analysis
RMSE_rssi_x = RMSE_rssi_x + (X_T - x)^2 ;
RMSE_rssi_y = RMSE_rssi_y+  (Y_T - y)^2;
RMSE_grnn_x = RMSE_grnn_x + (GRNN_Estimated_Loc(1) - x)^2 ;
RMSE_grnn_y = RMSE_grnn_y + (GRNN_Estimated_Loc(2) - y)^2;
RMSE_kf_x = RMSE_kf_x + (X_kalman - x)^2 ;
RMSE_kf_y = RMSE_kf_y + (Y_kalman - y)^2;
RMSE_ukf_x = RMSE_ukf_x + (X_ukf(1)-x)^2 ;
RMSE_ukf_y = RMSE_ukf_y + (X_ukf(2)-y)^2;
```

```
% ---> Part 2 : Calculation of Absolute Errors

% a) For Kalman Filter
error_x_kf(t) =  abs((x - X_kalman));
error_y_kf(t) =  abs((y - Y_kalman));
error_xy_kf(t) = ((error_x_kf(t) + error_y_kf(t))/2);


% b) For Unscented Kalman Filter
error_x_ukf(t) = abs((x - X_ukf(1)));
error_y_ukf(t) = abs((y - X_ukf(2)));
error_xy_ukf(t) = ((error_x_ukf(t) + error_y_ukf(t))/2);


% c) For Pure RSSI based Technique
error_x_rssi(t) = abs((x - X_T));
error_y_rssi(t) = abs((y - Y_T));
error_xy_rssi(t) =((error_x_rssi(t) + error_y_rssi(t))/2);


% d) For GRNN based Estimation
error_x_grnn(t) = abs((x - GRNN_Estimated_Loc(1)));
error_y_grnn(t) = abs((y - GRNN_Estimated_Loc(2)));
error_xy_grnn(t) =((error_x_grnn(t) + error_y_grnn(t))/2);

end

% Average Error in x &y  coordinates
avg_error_xy_rssi = 0; avg_error_xy_kf = 0 ;
avg_error_xy_grnn = 0; avg_error_xy_ukf = 0 ;

for t = 1:no_of_positions
avg_error_xy_rssi=avg_error_xy_rssi + (error_xy_rssi(t)/no_of_positions);
avg_error_xy_grnn=avg_error_xy_grnn + (error_xy_grnn(t)/no_of_positions);
avg_error_xy_kf= avg_error_xy_kf + (error_xy_kf(t)/no_of_positions);
avg_error_xy_ukf= avg_error_xy_ukf + (error_xy_ukf(t)/no_of_positions);
end

disp('Average Localization Errors :')
avg_error_xy_rssi
avg_error_xy_grnn
avg_error_xy_kf
avg_error_xy_ukf

disp('RMSE Errors :')
RMSE_rssi_x = sqrt(RMSE_rssi_x/no_of_positions)
RMSE_rssi_y = sqrt(RMSE_rssi_y/no_of_positions)
RMSE_grnn_x = sqrt(RMSE_grnn_x/no_of_positions)
RMSE_grnn_y = sqrt(RMSE_grnn_y/no_of_positions)
RMSE_kf_x = sqrt(RMSE_kf_x/no_of_positions)
RMSE_kf_y = sqrt(RMSE_kf_y/no_of_positions)
RMSE_ukf_x = sqrt(RMSE_ukf_x/no_of_positions)
RMSE_ukf_y = sqrt(RMSE_ukf_y/no_of_positions)
```

```
RMSE_rssi_avg = (RMSE_rssi_x + RMSE_rssi_y)/2
RMSE_grnn_avg = (RMSE_grnn_x + RMSE_grnn_y)/2
RMSE_kf_avg = (RMSE_kf_x + RMSE_kf_y)/2
RMSE_ukf_avg = (RMSE_ukf_x + RMSE_ukf_y)/2

% Plotting Absolute Errors of KF & UKF based Tracking

f2 = figure(2);
for t =1:no_of_positions

plot(t,error_x_grnn(t),'g+','LineWidth',2)
plot(t,error_x_kf(t),'b+','LineWidth',2)
plot(t,error_x_ukf(t),'k+','LineWidth',2)
plot(t,error_x_rssi(t),'ro','LineWidth',2)

xlabel('Time [in sec]','FontName','Times','Fontsize',14,'LineWidth',2)
ylabel('Error in x estimates [in meters]','FontName','Times','Fontsize',14,'LineWidth',2)
holdon

end
 legend('Error with Trilateration based Estimates','Error with GRNN based Estimates','Error with
GRNN+KF based Estimates','Error with GRNN+UKF based Estimates','Location','NorthWest')

 f3 = figure(3);
for t =1:no_of_positions

plot(t,error_y_grnn(t),'g+','LineWidth',2)
plot(t,error_y_kf(t),'b+','LineWidth',2)     %,'Markersize',2,'MarkerEdgeColor','b')
plot(t,error_y_ukf(t),'k+','LineWidth',2)    %,'Markersize',2,'MarkerEdgeColor','r')
plot(t,error_y_rssi(t),'ro','LineWidth',2)   %,'Markersize',2,'MarkerEdgeColor','g')

xlabel('Time [in sec]','FontName','Times','Fontsize',14, 'LineWidth',2)
ylabel('Error in y estimates [in meters]','FontName','Times','Fontsize',14, 'LineWidth',2)

holdon
end
 legend('Error with Trilateration based Estimates','Error with GRNN based Estimates','Error with
GRNN+KF based Estimates','Error with GRNN+UKF based Estimates','Location','NorthWest')

 f4 = figure(4);
for t =1:no_of_positions

plot(t,error_xy_grnn(t),'g+','LineWidth',2)
plot(t,error_xy_kf(t),'b+','LineWidth',2)     %,'Markersize',2,'MarkerEdgeColor','b')
plot(t,error_xy_ukf(t),'k+','LineWidth',2)    %,'Markersize',2,'MarkerEdgeColor','r')
plot(t,error_xy_rssi(t),'ro','LineWidth',2)   %,'Markersize',2,'MarkerEdgeColor','g')

xlabel('Time [in sec]','FontName','Times','Fontsize',14,'LineWidth',2)
ylabel('Error in xy estimates [in meters]','FontName','Times','Fontsize',14,'LineWidth',2)

holdon
end
 legend('Error with Trilateration based Estimates','Error with GRNN based Estimates','Error with
GRNN+KF based Estimates','Error with GRNN+UKF based Estimates','Location','NorthWest')
```

MATLAB Code for Case III

```
%%%%%%%%%%%%%%% main.m %%%%%%%%%%%%%%%%%

%% WSN Deployment Setting Parameters
clearall
closeall
clc

   N = 4;                          % number of anchors
   M = 1;                          % number of mobile nodes
networkSize = 100;                   % we consider a 100by100 area that the mobile can
wander

anchorLoc   = [0              0;        % set the anchor at 4 vertices of the region
networkSize        0;
           0        networkSize;
networkSizenetworkSize];

%%% GRNN Setting %%%%%%%%%%%%%%%%%%%%%%%%%%%%%%%

RSSI_input_Vector = [-8.63711087700203,-11.5121738440855,-14.9788577741561,-
18.2424970326168,-17.2767505383621,-22.1529456887463,-20.5988132710408,-
26.7381951841692,-28.1016056626748,-26.8401681592955,-31.2009622756918,-
30.7067924858325,-27.0108637314813,-28.6009147429885,-34.8669185548126,-
33.6713973358156,-26.1945007053475,-34.1098185486167,-32.7378316422370,-
29.5555232963034,-29.1302982767054,-26.0425571563218,-31.0142194231226,-
23.9517855153775,-35.3809753208000,-30.8689220924318,-23.2519641062010,-
26.3382137588999,-33.2003851315704,-26.6366098384704,-26.8893597649793,-
28.1481934702383,-29.3271622723670,-29.3502853536700,-22.2650755453830,-
32.9478086658446,-19.1908979492906,-26.5517429750970,-27.8882690706260,-
24.0805034710059;-24.6443378867791,-23.8593396704728,-25.0300774444898,-
27.3552769837086,-31.3238874955404,-25.5684451777475,-27.2669962899477,-
28.5236948407618,-30.6095309413261,-27.3550681549064,-29.0951633851164,-
26.8496351417526,-25.2402279674739,-33.4366094367594,-28.2018648304488,-
31.7202588327514,-25.7273993271116,-27.7406442000865,-26.8123711146280,-
25.8920330826360,-32.8645336994140,-22.4362331219866,-22.7408893670739,-
26.6055760301610,-15.1061425078805,-22.8393577737386,-16.5018015844959,-
20.3573928149789,-25.9018275788868,-19.0747887851302,-16.4556233464885,-
17.9921345751942,-18.2818096318861,-18.0160361661470,-21.0946978981921,-
14.8199088088597,-17.4389955511778,-14.5491119152686,-16.9372270707295,-
21.7719689162081;-28.5135613143206,-33.0780989357883,-31.3506779461191,-
36.0066734173278,-36.4771716278969,-26.2060894708195,-31.9861022431075,-
25.0584109759550,-28.0201092305810,-20.7125177645295,-20.6950504734078,-
20.6937065669968,-25.6230277199468,-18.7990725652173,-19.3405715045320,-
13.7283029756786,-15.7661646303376,-16.9170765891923,-12.4096156327949,-
20.4488014567320,-11.2793351657313,-19.2735444611160,-12.9127466593933,-
```

```
16.6973307478935,-23.2440861614554,-19.5423962483160,-17.8774561543769,-
22.2394269412980,-23.2321402317115,-18.2863155370521,-22.0255637228001,-
25.5137093113793,-26.9308086311955,-21.8929688622825,-19.7266313616989,-
32.6206155831883,-24.9869109502726,-28.0621473668449,-28.2854211807112,-
32.7967125887547;-25.9118064691782,-27.2885447453488,-23.5263426077213,-
19.7371525265260,-17.6956931946208,-20.1876598879298,-18.2350285824841,-
18.2568318614458,-16.4786757425207,-23.8035181257022,-23.5704645802868,-
19.9652392972093,-21.9046157843161,-19.8548782223887,-21.6613529489605,-
22.3355342877953,-23.4218741164849,-22.8986277147529,-26.6970793451071,-
24.5674064740980,-22.1249165070594,-30.4306177007705,-30.4071198812238,-
28.1721577460586,-30.6151353968228,-28.0296173820039,-31.5003675294334,-
28.3877986063688,-33.0397705487563,-32.0049067239378,-33.2699125292289,-
31.2200071293077,-33.9181605899042,-31.9872995878781,-29.6661170371290,-
29.1643624501848,-28.1690529175130,-27.5642192144777,-32.5581354439923,-
34.6179330646867];
Target =
[10,13,16,19,22,25,28,31,34,38,42,46,50,54,58,62,62,62,64,66,68,70,72,74,76,78,80,82,84,86,84
,82,80,78,76,74,72,70,68,66;10,17,24,31,38,45,52,59,66,68,70,72,74,76,78,80,80,80,77,74,71,68
,65,62,59,56,53,50,47,44,42,40,38,36,34,32,30,28,26,24];
spread =3.5;

net_Loc_est = newgrnn(RSSI_input_Vector,Target,spread);
view(net_Loc_est)

%show anchor Locations
  f1 = figure(1);
  plot(anchorLoc(:,1),anchorLoc(:,2),'ko','MarkerSize',8,'lineWidth',2,'MarkerFaceColor','k');
axis([0 100 0 100])
gridon
holdon

% Definingveriables
no_of_positions = 35;    % Total Simulation Period = 35 seconds


  P = [0.25 0 0  0; 0 0.04 0 0; 0 0 0.02 0; 0 0 0 0.01];  % Initial Process Covariance Matrix
  q=0.1;                                   %std of process

%%%%%%%%%%%%%% Initialization of Error Values %%%%%%%%%%%%%%%%%
  RMSE_kf1_x =0 ;RMSE_rssi_x= 0; RMSE_ukf1_x= 0; RMSE_grnn_x = 0;
  RMSE_kf1_y =0 ;RMSE_rssi_y= 0; RMSE_ukf1_y= 0; RMSE_grnn_y = 0;
  RMSE_kf2_x =0 ;RMSE_rssi_x= 0; RMSE_ukf2_x= 0; RMSE_grnn_x = 0;
  RMSE_kf2_y =0 ;RMSE_rssi_y= 0; RMSE_ukf2_y= 0; RMSE_grnn_y = 0;

% Calculate reference RSSI at d0 = 1 meter using Free Space Path Loss Model
  d0=1;                                    % in Meters
```

```
    Pr0 = RSSI_friss(d0);

d_test = 20;

Pr = RSSI_friss(d_test);

%Calculation of Path Loss Exponent :

    n = -(Pr + Pr0)/(10*log(d_test));
    x=10;
    y=10;

% Generating trajectory for the mobile node
for t = 1:no_of_positions
if(t<9)              x_v = 3; y_v = 7;
elseif(t==9 && t<16)  x_v = 4; y_v = 2;
elseif(t>=16 && t<18)  x_v = 0; y_v = 0;
elseif(t>=18 && t<30)  x_v = 2; y_v = -3;
elseif(t>=30 && t<=40) x_v = -2; y_v = -2;
end

x_actual(t)=x;      % para 1
y_actual(t)=y;      % para 1

    x=x+x_v;
    y=y+y_v;

disp('True Location')
    [x,y]                          % Actual Target Location
plot(x,y,'rs','LineWidth',2)

ylabel('y-Axis[meter]','FontName','Times','Fontsize',14,'LineWidth',2);
xlabel('x-Axis[meter]','FontName','Times','Fontsize',14,'LineWidth',2);
    legend('Anchor Node Location','Actual Target Track','GRNN+KF based
Estimation','GRNN+UKF based Estimation','Trilateration+KF based
Estimation','Trilateration+UKF Based Implementation','Location','SouthEast')
holdon

% Actual Distances from Anchors required to generate RSSI Values
    d1 = sqrt( x^2 + y^2 );
    d2 = sqrt((100-x)^2 + y^2);
    d3 = sqrt((100-x)^2+ (100-y)^2);
    d4 = sqrt(x^2+ (100-y)^2);

% Generate RSSI Values at 4 Anchor Nodes which are at d1, d2, d3 & d4 distances respectively
from Moving Target

    RSS = lognormalshadowing_4(n,d1,d2,d3,d4,Pr0);

    RSS_1(t)= RSS(1);
    RSS_2(t)= RSS(2);
    RSS_3(t)= RSS(3);                %4 RSS Values
    RSS_4(t)= RSS(4);
```

```
RSS_new_vector = RSS.';
GRNN_Estimated_Loc = sim(net_Loc_est ,RSS_new_vector)

GRNN_x(t)= GRNN_Estimated_Loc(1);
GRNN_y(t)= GRNN_Estimated_Loc(2);

    RSS_s = sort(RSS);

disp('RSSI Estimated Location')

mobileLoc_est = trilateration_4(RSS,RSS_s,Pr0,n,networkSize)

    X_T = mobileLoc_est(1);
    Y_T = mobileLoc_est(2);

trad_x(t)=X_T;
trad_y(t)=Y_T;

%plot(X_T,Y_T,'r+','LineWidth',2)
%hold on

% calculate velocities in X & Y Directions

velocity_est=velocity(X_T,Y_T,t);

      s = [x; y; 0; 0];                    % State of the system at time 't'
      X = s;                      % state with noise

% GRNN + Kalman Filter for Tracking Moving Target Code starts here
    [X_kalman1,Y_kalman1,X_kf]= kf(X,P,GRNN_Estimated_Loc(1),
GRNN_Estimated_Loc(2), velocity_est,t);
    kf1_x(t)=X_kalman1;     %Para 3
    kf1_y(t)=Y_kalman1;     %Para 4
plot(X_kalman1,Y_kalman1,'ko','LineWidth',2)
holdon

%GRNN + UKF implementation
    Z = [GRNN_Estimated_Loc(1); GRNN_Estimated_Loc(2); velocity_est(1); velocity_est(2)];

disp('UKF Estimated State')
    [X_ukf1]= ukf5(X,P,Z)
    ukf1_x(t)=X_ukf1(1);     %Para 5
    ukf1_y(t)=X_ukf1(2);     %Para 6
plot(X_ukf1(1),X_ukf1(2),'mo','LineWidth',2)
holdon

% Trad. RSSI + Kalman Filter for Tracking Moving Target Code starts here
    [X_kalman2,Y_kalman2,X_kf]= kf(X,P,X_T, Y_T, velocity_est,t);
    kf2_x(t)=X_kalman2;     %Para 3
    kf2_y(t)=Y_kalman2;     %Para 4
plot(X_kalman2,Y_kalman2,'g+','LineWidth',2)
holdon
```

```
%Trad. RSSI + UKF implementation
    Z = [X_T; Y_T; velocity_est(1); velocity_est(2)];
disp('UKF Estimated State')
    [X_ukf2]= ukf5(X,P,Z)
    ukf2_x(t)=X_ukf2(1);     %Para 5
    ukf2_y(t)=X_ukf2(2);     %Para 6
plot(X_ukf2(1),X_ukf2(2),'b+','LineWidth',2)
holdon


%%%%% Error Analysis of algorithm
% ---> Part 1 : RMSE Analysis

     RMSE_kf1_x = RMSE_kf1_x + (X_kalman1 - x)^2 ;
     RMSE_kf1_y = RMSE_kf1_y + (Y_kalman1 - y)^2;
     RMSE_kf2_x = RMSE_kf2_x + (X_kalman2 - x)^2 ;
     RMSE_kf2_y = RMSE_kf2_y + (Y_kalman2 - y)^2;

     RMSE_ukf1_x = RMSE_ukf1_x + (X_ukf1(1)-x)^2 ;
     RMSE_ukf1_y = RMSE_ukf1_y + (X_ukf1(2)-y)^2;
     RMSE_ukf2_x = RMSE_ukf2_x + (X_ukf2(1)-x)^2 ;
     RMSE_ukf2_y = RMSE_ukf2_y + (X_ukf2(2)-y)^2;

% ---> Part 2 : Calculation of Absolute Errors

% a) For Kalman Filter
     error_x_kf1(t) =  abs((x - X_kalman1));
     error_y_kf1(t) =  abs((y - Y_kalman1));
     error_xy_kf1(t) = ((error_x_kf1(t) + error_y_kf1(t))/2);
     error_x_kf2(t) =  abs((x - X_kalman2));
     error_y_kf2(t) =  abs((y - Y_kalman2));
     error_xy_kf2(t) = ((error_x_kf2(t) + error_y_kf2(t))/2);

% b) For Unscented Kalman Filter
     error_x_ukf1(t) = abs((x - X_ukf1(1)));
     error_y_ukf1(t) = abs((y - X_ukf1(2)));
     error_xy_ukf1(t) = ((error_x_ukf1(t) + error_y_ukf1(t))/2);
     error_x_ukf2(t) = abs((x - X_ukf2(1)));
     error_y_ukf2(t) = abs((y - X_ukf2(2)));
     error_xy_ukf2(t) = ((error_x_ukf2(t) + error_y_ukf2(t))/2);

end

% Average Error in x &y  coordinates
% avg_error_xy_rssi = 0; avg_error_xy_grnn = 0;
    avg_error_xy_kf1 = 0 ;avg_error_xy_kf2 = 0 ;
    avg_error_xy_ukf1 = 0 ;avg_error_xy_ukf2 = 0 ;
```

```
for t = 1:no_of_positions

    avg_error_xy_kf1= avg_error_xy_kf1 + (error_xy_kf1(t)/no_of_positions);
    avg_error_xy_ukf1= avg_error_xy_ukf1 + (error_xy_ukf1(t)/no_of_positions);
    avg_error_xy_kf2= avg_error_xy_kf2 + (error_xy_kf2(t)/no_of_positions);
    avg_error_xy_ukf2= avg_error_xy_ukf2 + (error_xy_ukf2(t)/no_of_positions);
end

disp('Average Localization Errors :')
%avg_error_xy_rssi
%avg_error_xy_grnn
   avg_error_xy_kf1
   avg_error_xy_ukf1
   avg_error_xy_kf2
   avg_error_xy_ukf2

disp('RMSE Errors :')

   RMSE_kf1_x = sqrt(RMSE_kf1_x/no_of_positions)
   RMSE_kf1_y = sqrt(RMSE_kf1_y/no_of_positions)
   RMSE_ukf1_x = sqrt(RMSE_ukf1_x/no_of_positions)
   RMSE_ukf1_y = sqrt(RMSE_ukf1_y/no_of_positions)
   RMSE_kf2_x = sqrt(RMSE_kf2_x/no_of_positions)
   RMSE_kf2_y = sqrt(RMSE_kf2_y/no_of_positions)
   RMSE_ukf2_x = sqrt(RMSE_ukf2_x/no_of_positions)
   RMSE_ukf2_y = sqrt(RMSE_ukf2_y/no_of_positions)


    RMSE_kf1_avg = (RMSE_kf1_x + RMSE_kf1_y)/2
    RMSE_ukf1_avg = (RMSE_ukf1_x + RMSE_ukf1_y)/2
    RMSE_kf2_avg = (RMSE_kf2_x + RMSE_kf2_y)/2
    RMSE_ukf2_avg = (RMSE_ukf2_x + RMSE_ukf2_y)/2

% Plotting Absolute Errors of KF & UKF based Tracking

f2 = figure(2);
for t =1:no_of_positions

plot(t,error_x_ukf1(t),'ro','LineWidth',2)
plot(t,error_x_kf2(t),'g+','LineWidth',2)
plot(t,error_x_ukf2(t),'b+','LineWidth',2)
plot(t,error_x_kf1(t),'ko','LineWidth',2)

xlabel('Time [in sec]','FontName','Times','Fontsize',14,'LineWidth',2)
ylabel('Error in x estimates [in meter]','FontName','Times','Fontsize',14,'LineWidth',2)
holdon

end
  legend('Error with GRNN+KF based Estimation','Error with GRNN+UKF based
Estimation','Error with Trilateration+KF based Estimation','Error with Trilateration+UKF based
Estimation','Location','NorthWest')
```

```
  f3 = figure(3);
for t =1:no_of_positions


plot(t,error_y_ukf1(t),'ro','LineWidth',2)    %,'Markersize',2,'MarkerEdgeColor','r')
plot(t,error_y_kf2(t),'g+','LineWidth',2)     %,'Markersize',2,'MarkerEdgeColor','b')
plot(t,error_y_ukf2(t),'b+','LineWidth',2)    %,'Markersize',2,'MarkerEdgeColor','r')
plot(t,error_y_kf1(t),'ko','LineWidth',2)     %,'Markersize',2,'MarkerEdgeColor','b')
xlabel('Time [in sec]','FontName','Times','Fontsize',14, 'LineWidth',2)
ylabel('Error in y estimates [in meter]','FontName','Times','Fontsize',14, 'LineWidth',2)
holdon
end
  legend('Error with GRNN+KF based Estimation','Error with GRNN+UKF based
Estimation','Error with Trilateration+KF based Estimation','Error with Trilateration+UKF based
Estimation','Location','NorthWest')

  f4 = figure(4);
for t =1:no_of_positions

plot(t,error_xy_ukf1(t),'ro','LineWidth',2)    %,'Markersize',2,'MarkerEdgeColor','r')
plot(t,error_xy_kf2(t),'g+','LineWidth',2)     %,'Markersize',2,'MarkerEdgeColor','b')
plot(t,error_xy_ukf2(t),'b+','LineWidth',2)    %,'Markersize',2,'MarkerEdgeColor','r')
plot(t,error_xy_kf1(t),'ko','LineWidth',2)     %,'Markersize',2,'MarkerEdgeColor','b')
xlabel('Time [in sec]','FontName','Times','Fontsize',14,'LineWidth',2)
ylabel('Error in xy estimates [in meter]','FontName','Times','Fontsize',14,'LineWidth',2)
holdon
end
  legend('Error with GRNN+KF based Estimation','Error with GRNN+UKF based
Estimation','Error with Trilateration+KF based Estimation','Error with Trilateration+UKF based
Estimation','Location','NorthWest')
```

References

1. S. K. Gharghan, R. Nordin, M. Ismail, J. A. Ali, Accurate wireless sensor localization technique based on hybrid pso-ann algorithm for indoor and outdoor track cycling. IEEE Senors J. (2016). https://doi.org/10.1109/JSEN.2015.2483745
2. F. Viani, P. Rocca, G. Oliveri, D. Trinchero, A. Massa, Localization, tracking, and imaging of targets in wireless sensor networks: an invited review. Radio Sci. (2011). https://doi.org/10.1029/2010RS004561
3. H. Huang, L. Chen, E. Hu, A neural network-based multi-zone modelling approach for predictive control system design in commercial buildings. Energy Build. (2015). https://doi.org/10.1016/j.enbuild.2015.03.045
4. S.R. Jondhale, R.S. Deshpande, Kalman filtering framework based real time target tracking in wireless sensor networks using generalized regression neural networks. IEEE Sensors J. **19**, 224–233 (2018)
5. S. R. Jondhale, R. S. Deshpande, GRNN and KF framework based real time target tracking using PSOC BLE and smartphone. Ad Hoc Netw. (2019). https://doi.org/10.1016/j.adhoc.2018.09.017

6. S. Jondhale, R. Deshpande, Self recurrent neural network based target tracking in wireless sensor network using state observer. Int. J. Sensors Wirel. Commun. Control (2018). https://doi.org/10.2174/2210327908666181029103202
7. S.R. Jondhale, R.S. Deshpande, Efficient localization of target in large scale farmland using generalized regression neural network. Int. J. Commun. Syst. **32**(16), e4120 (2019). https://doi.org/10.1002/dac.4120
8. D. Bani-Hani, M. Khasawneh, A recursive general regression neural network (R-GRNN) Oracle for classification problems. Expert Syst. Appl. **135**, 273–286 (2019). https://doi.org/10.1016/j.eswa.2019.06.018
9. D. F. Specht, Probabilistic neural networks. Neural Netw. (1990). https://doi.org/10.1016/0893-6080(90)90049-Q
10. D. F. Specht, A general regression neural network. IEEE Trans. Neural Netw. (1991). https://doi.org/10.1109/72.97934
11. D. F. Specht, GRNN with double clustering, in *The 2006 IEEE International Joint Conference on Neural Network Proceedings* (2006). https://doi.org/10.1109/ijcnn.2006.247235
12. Q. Wen, P. Qicong, An improved particle filter algorithm based on neural network, in *Intelligent Information Processing III. IIP 2006. IFIP International Federation for Information Processing*, (Springer, Boston, 2006)
13. S.R. Jondhale, R.S. Deshpande, GRNN and KF framework based real time target tracking using PSOC BLE and smartphone. Ad Hoc Netw. **84**, 19–28 (2019). https://doi.org/10.1016/j.adhoc.2018.09.017
14. N. Patwari, J. N. Ash, S. Kyperountas, A. O. Hero, R. L. Moses, N. S. Correal, Locating the nodes: cooperative localization in wireless sensor networks. IEEE Signal Process. Mag. (2005). https://doi.org/10.1109/MSP.2005.1458287
15. A. PAL, Localization algorithms in wireless sensor networks: current approaches and future challenges. Netw. Protoc. Algorithms (2011). https://doi.org/10.5296/npa.v2i1.279
16. L. Gogolak, S. Pletl, D. Kukolj, Neural network-based indoor localization in WSN environments. Acta Polytech. Hungarica **10**, 221–235 (2013)
17. S.R. Jondhale, R.S. Deshpande, Modified Kalman filtering framework based real time target tracking against environmental dynamicity in wireless sensor networks. Ad Hoc Sens. Wirel. Netw. **40**, 119–143 (2018)

Chapter 7
Supervised Learning Architecture-Based L&T Using RSSI

7.1 Supervised Learning Architectures for L&T

7.1.1 FFNT

From the name it is clear that in FFNT the data moves to the output nodes in forward direction only [1–3]. The FFNT architecture includes three layers: input, hidden, and output layers (see Fig. 7.1) [4–7]. The input layer is supplied with vector$X = [RSSI1, RSSI2, RSSI3, RSSI4]$.

The FFNT yields the estimated location $(\hat{x},\hat{y})$ using Eq. (7.1).

$$(\hat{x},\hat{y}) = \sum_{i=1}^{H} \mathbf{W}_{\mathbf{i}}^{(2)} \sigma \left(\sum_{j=1}^{N} \mathbf{W}_{\mathbf{ij}}^{(1)} X(i) + \mathbf{b}_{\mathbf{i}} \right) \tag{7.1}$$

where:

$\mathbf{W}_{\mathbf{i}}^{(1)}$ – interconnection weight between input node j and hidden node i.

$\mathbf{W}_{\mathbf{i}}^{(2)}$ – interconnection weight between hidden node i and output node.

b_{i} – bias for hidden node i.

7.1.2 Radial Basis Function Neural Network (RBFN or RBFNN)

The RBFNN network is popular for hidden function approximation of any continuous function [8–10]. It is capable to converge at fast rate compared to other ANN architectures. It is basically a three-layered FFNT (see Fig. 7.2). As we are

S. R. Jondhale et al., *Received Signal Strength Based Target Localization and Tracking Using Wireless Sensor Networks*, EAI/Springer Innovations in Communication and Computing, https://doi.org/10.1007/978-3-030-74061-0_7

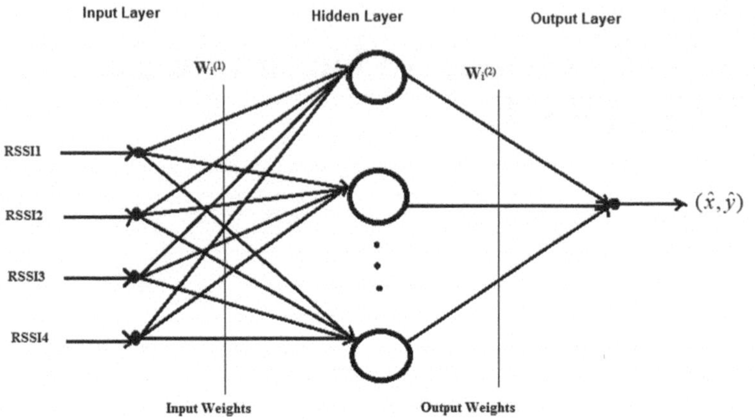

Fig. 7.1 The FFNT-based architecture for L&T

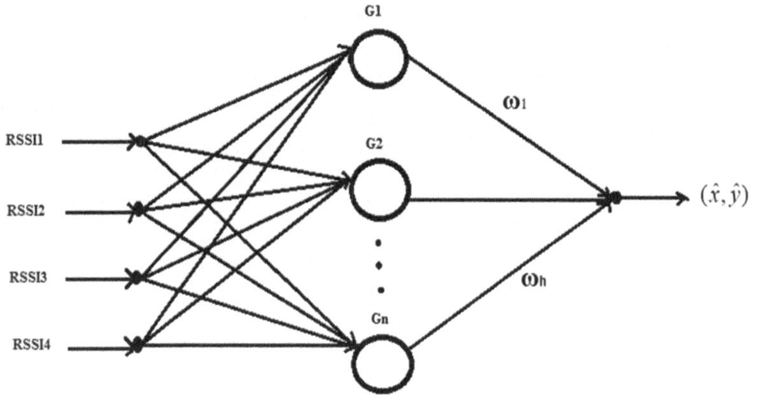

Fig. 7.2 The RBFNN architecture for L&T

discussing the application of the RBFNN in the RSSI-based target L&T domain, we are applying RSSI measurements at the input of the network, and we get the estimated target location at the output terminal of the network as shown in Fig. 7.1. The mapping between input and hidden layer is generally nonlinear and is achieved using RBF function. The mapping between the hidden layer and output layer is generally a linear function.

The RBFNN works in two steps. In the first step the center vector and width parameters are to be determined using applied input samples. Thus, the first step is unsupervised learning step. Once the hidden layer parameters are computed, the interconnection weights between the hidden layer and output layer are determined using the simple least squares method in the second step. A wide variety of radial basis functions are possible for the hidden layer. In this chapter we are going to use the popular Gaussian function. The detailed mathematical expression for RBFNN-based estimation is given in Eq. (7.2) [8–10]:

$$G(X-c_i)=\exp\left(-\frac{1}{2\sigma_i^2}\|X-c_i\|^2\right) \tag{7.2}$$

where:

X—state vector consisting of RSSI measurements $X = [RSSI1, RSSI2, RSSI3, RSSI4]$.

$\|X - c_i\|$—Euclidean distance.

c_i—central vector of the Gaussian function.

$w_i(i = 1,2,\ldots h)$—interconnection weight between hidden layer and output layer.

(x, y)—output of RBFNN network.

The estimated location of the target using RBFNN is given in Eq. (7.3) [8–10]:

$$(\hat{x},\hat{y})=\sum_{i=1}^{h} w_h \exp\left(-\frac{1}{2\sigma_i^2}\|X-c_i\|^2\right) \tag{7.3}$$

7.1.3 Multilayer Perceptron (MLP)

The MLP architecture is similar to the FFNN architecture with three layers as illustrated in Fig. 7.3. The activation function for hidden layer is piecewise linear in nature [11–14]. Let's discuss MLP parameters and their notations.

$\mathbf{W}_i^{(1)}(i = 1,2,\ldots,N)$—interconnection weight between input layer to hidden layer.

$\mathbf{W}_i^{(2)}(i = 1,2,\ldots,N)$—interconnection weight between hidden layer to output layer.

$b_i^{(1)}(i = 1,2,\ldots,N)$—biases for hidden layer nodes.

$b_i^{(2)}(i = 1,2,\ldots,N)$—biases for output layer nodes.

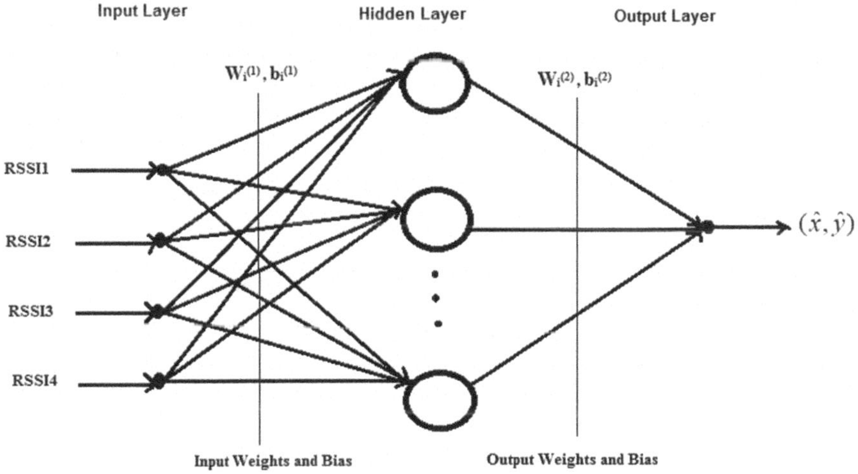

Fig. 7.3 MLP network for L&T

The estimated location $(\hat{x},\hat{y})$ obtained using MLP network is given in Eq. (7.4):

$$(\hat{x},\hat{y}) = \sum_{i=1}^{N} \mathbf{W}_{\mathrm{i}}^{(2)} \varphi\left(\mathbf{W}_{\mathrm{i}}^{(1)} X + \mathbf{b}_{\mathrm{i}}^{(1)}\right) + \mathbf{b}_{\mathrm{i}}^{(2)} \tag{7.4}$$

where $X = [RSSI1, RSSI2, RSSI3, RSSI4]$.

7.2 Training Functions in ANN

A training function in ANN defines how the interconnection weights are updated in the network after iterations of ANN. A wide variety of training functions are available for ANN training [15–20]. The detailed list is given in Table 7.1. Choosing appropriate training function is very crucial for the given underlying application. For target L&T problem considered in this chapter, all of these training functions are tested with proposed ANN architectures.

7.3 Application of Supervised Learning Architectures for L&T

This section discusses the applications of FFNT, RBFNN, and MLP architectures to solve the problem of the moving target L&T. All of these architectures will be trained with a training dataset in off-line step. This dataset includes 30 sets of 2-D locations of the target and corresponding four RSSI measurements. Each set is one input vector with four RSSI measurements. In order to compare and verify all proposed supervised learning architectures simultaneously, all of them are fed with the

Table 7.1 Various ANN training functions with their group number

Sr. no.	Acronym	Description
1	GD	Gradient descent backpropagation
2	GDX	Gradient descent with momentum and adaptive learning rate backpropagation
3	GDA	Gradient descent with adaptive learning rate backpropagation
4	RP	Resilient backpropagation
5	CGP	Conjugate gradient backpropagation with Polak/Ribiére restarts
6	CGF	Conjugate gradient backpropagation with fletcher-reeves restarts
7	CGB	Conjugate gradient with Powell/Beale restarts
8	BFG	BFG quasi-Newton backpropagation
9	SCG	Scaled conjugate gradient backpropagation
10	LM	Levenberg-Marquardt backpropagation
11	OSS	One-step secant backpropagation

same four random RSSI measurements online, and they are supposed to estimate 2-D target location, corresponding to this input vector of four RSSI measurements. These architectures are tested against noise uncertainty in RSSI measurements, arising out of issues related with signal propagation such as NLOS, multipath fading, and reflections. Once trained off-line the proposed architectures are tested with 35 sets of input vectors, corresponding to 35 target locations during target motion. To assess the performance of the proposed ANN architectures against fluctuations or variations in RSSI measurements, the RSSI noise variance is changed from 0 dBm to 5 dBm in steps of 1 dBm. The details of the considered target L&T problem are discussed below.

7.3.1 System Assumptions and Design

The indoor area 100 meters × 100 meters is simulated using MATLAB 2013a. The unknown target locations and anchor nodes are deployed in this area and are depicted by red and blue markers in Fig. 7.4. The anchor density and unknown target locations considered in this study are 4 and 24, respectively. The details of deployment of the target locations, anchor nodes, and non-anchor nodes are given in Tables 7.2 and 7.3. The communication radius and transmitted power in the

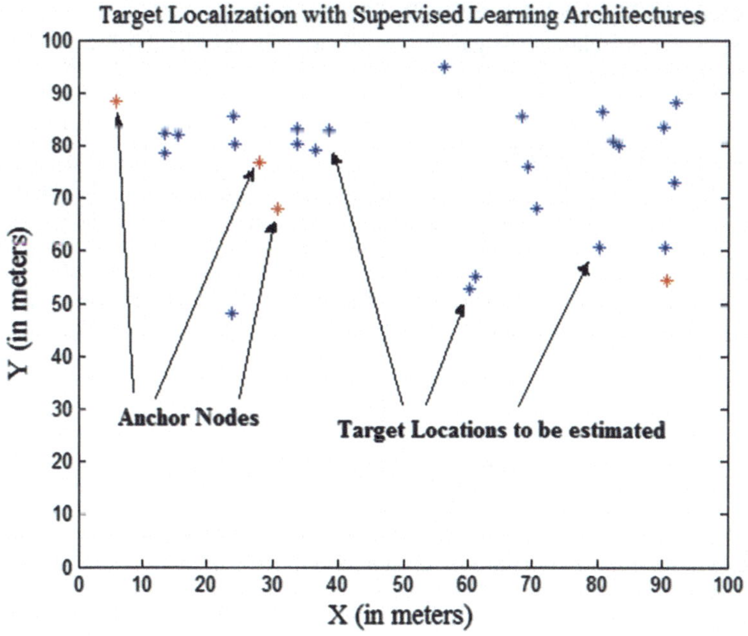

Fig. 7.4 Deployment of anchor nodes and target locations to be estimated for Case I and Case II

Table 7.2 Deployments of anchor nodes with 2-D locations for Case I and Case II

Anchor node number	2-D location	Anchor node number	2-D location
1	(5.93, 88.50)	3	(30.68, 68.03)
2	(27.92,76.89)	4	(90.75, 54.45)

Table 7.3 Unknown target locations to be estimated in Case I and Case II

Unknown location number	Unknown location of target to be estimated	Unknown location number	Unknown location of target to be estimated
1	(23.86, 48.17)	13	(68.27, 85.39)
2	(69.16, 75.90)	14	(80.21, 60.70)
3	(70.52, 68.05)	15	(24.24, 80.14)
4	(91.95, 88.10)	16	(15.57, 81.90)
5	(24.09, 85.62)	17	(33.81, 83.24)
6	(56.26, 94.75)	18	(82.40, 80.80)
7	(80.72, 86.36)	19	(90.22, 83.51)
8	(91.83, 72.96)	20	(33.79, 80.30)
9	(38.53, 82.95)	21	(60.28, 52.66)
10	(61.09, 55.00)	22	(83.18, 80.03)
11	(13.48, 78.54)	23	(36.41, 79.04)
12	(13.53, 82.42)	24	(90.45, 60.71)

simulation experiments are set to 0 dBm and 30 m, respectively. This research work follows the LNSM model to generate RSSI measurements.

This chapter presents investigation on two cases as described below:

- *Case I: Performance evaluation of FFNT with various ANN training functions against the traditional trilateration technique in the context of target L&T with variation in RSSI measurement noise.*
- *Case II: Comparison of the trilateration, GRNN, RBFNN, FFNT, and MLP architectures in the context of target L&T (Fig. 7.5).*

where:

X_i $(i = 1, 2, \ldots, p)$ is the training set containing 30 RSSI vectors with noise variance in RSSI measurement of 0 dBm and corresponding p unknown locations (off-line training step).

$$p = 24.$$

$$X_i = \left[RSSI1, RSSI2, RSSI3, RSSI4\right].$$

During the location estimation step (online estimation step), the noise variance is changed gradually from 0 dBm to 5 dBm as discussed earlier.

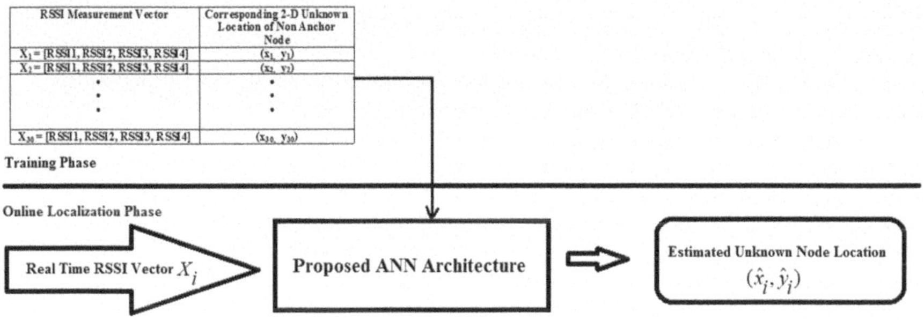

Fig. 7.5 System design of the supervised learning-based target L&T (Case II)

7.3.2 Evaluation Parameters

In order for the performance validation of the proposed supervised learning architectures, the following two parameters are utilized (see Eqs. (7.5) and (7.6)) [21–24]. The detailed discussion is made on these parameters in the previous chapters.

$$\text{Localization Error} = \sqrt{\left(x_i - \hat{x}_i\right)^2 + \left(y_i - \hat{y}_i\right)^2}, \quad i = 1,2,\ldots p. \tag{7.5}$$

$$\text{Average Localization Error} = \frac{\sum_{i=1}^{p} \text{Localization Error}}{p} \tag{7.6}$$

7.3.3 Algorithmic Flow of Proposed ANN Architectures

The flow of the proposed supervised learning algorithms for target L&T includes three steps (see Table 7.4).

7.3.4 Discussion on Results

- ***Case I**: Performance evaluation of FFNT with various ANN training functions against trilateration wrt variation in RSSI measurement noise.*

The simulation results for Case I are presented in Figs. 7.6, 7.7, 7.8, 7.9, 7.10, 7.11, and 7.12. In all of these results, the following color combinations are used:

Red marker "∗"—location of anchor nodes.

Blue marker "*"—actual location of unknown nodes.

Table 7.4 Flow of target localization for Case I and Case II

I. ANN training (off-line step) *Step 1*: Here FFNT is trained with 30 sets of RSSI measurements and corresponding actual target locations with the help of different training functions
II. Location estimation (online step) *Step 2*: • *For trilateration-based estimation:* Collect RSSIs transmitted by 12 anchor nodes and dispatch these to the base station • *For GRNN-, MLP-, RBFN-, and FFNT-based estimations:* Collect RSSIs transmitted by 12 anchor nodes and dispatch these to the base station *Step 3*: • *For trilateration-based estimation:* Execute trilateration to estimate the target location at the base station. Compute and record the localization errors along x and y directions using Eq. (7.5) • *For GRNN-, MLP-, RBFN-, and FFNT-based estimations:* Execute *GRNN, MLP, RBFN, and FFNT* algorithms to estimate the target location at the base station. Compute and record the localization errors along x and y directions using Eq. (7.5)
III. Computation of the average localization error *Step 4*: The average localization error for all the simulation experiments are computed using Eq. (6)

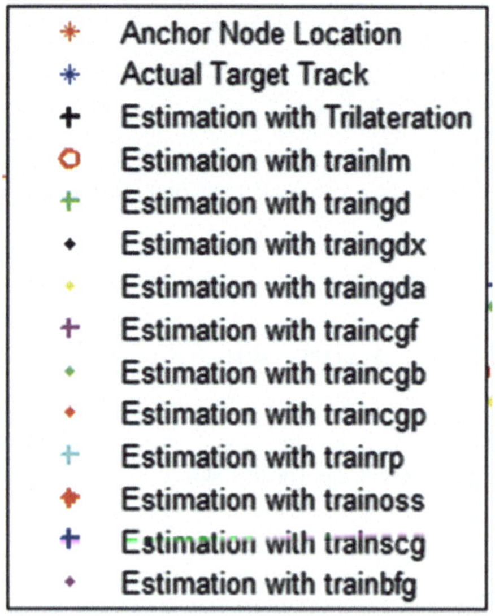

Fig. 7.6 Color combinations adopted in simulation results for L&T using trilateration and FNNN with various training functions (Case I and Case II)

Black marker "+"—estimated unknown location using trilateration.

The color combinations for showing the estimations with FFNT trained with various training functions are depicted in Fig. 7.6. The performance comparison of the proposed FFNT (network trained) and trilateration is depicted in Figs. 7.7, 7.8, 7.9, 7.10, 7.11, and 7.12 for all the RSSI noise variance cases. The numeric comparison results of the average localization errors are presented in Table 7.5.

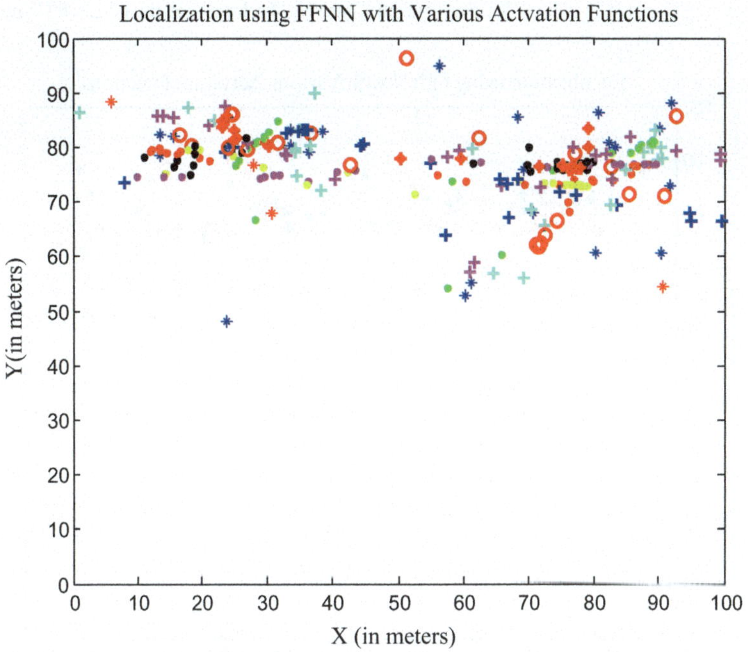

Fig. 7.7 Localization using trilateration and FNNN with various training functions with variance of measurement noise = 3 dBm (Case I)

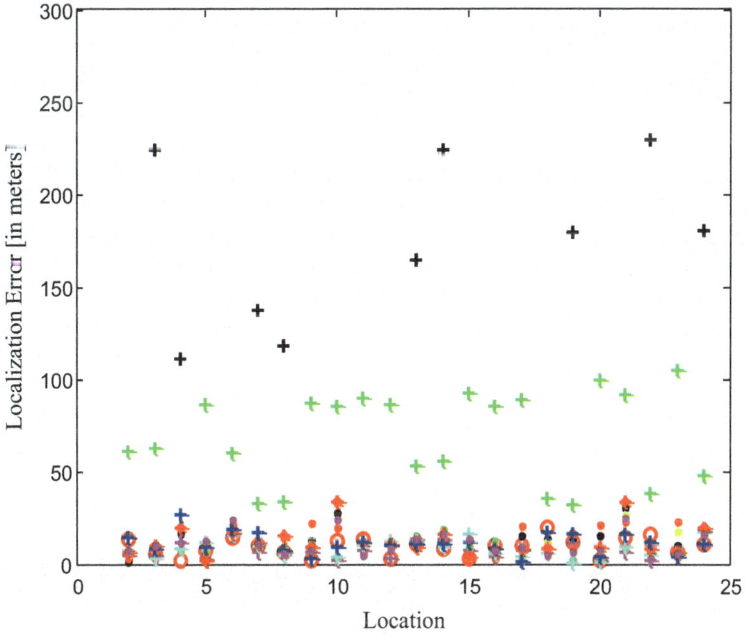

Fig. 7.8 Average localization error with trilateration and FNNN with various training functions with variance of measurement noise = 3 dBm (Case I)

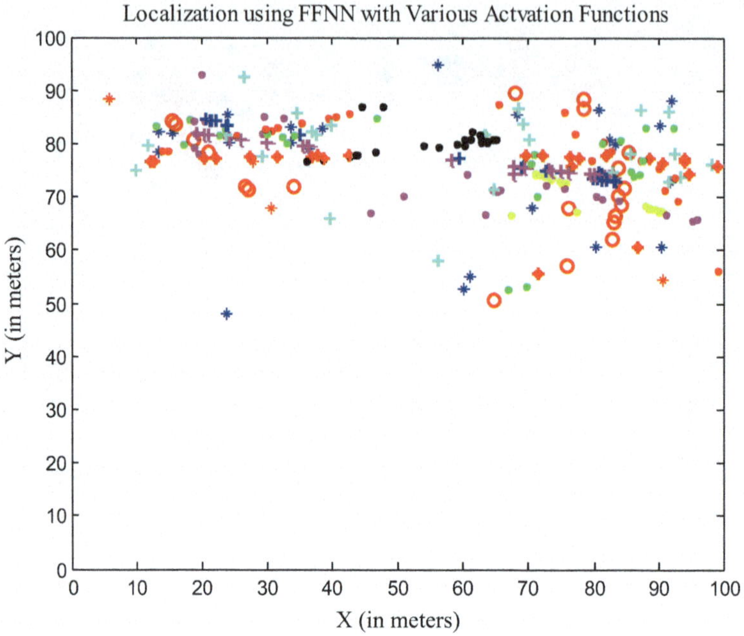

Fig. 7.9 Localization using trilateration and FNNN with various training functions with variance of measurement noise = 4 dBm (Case I)

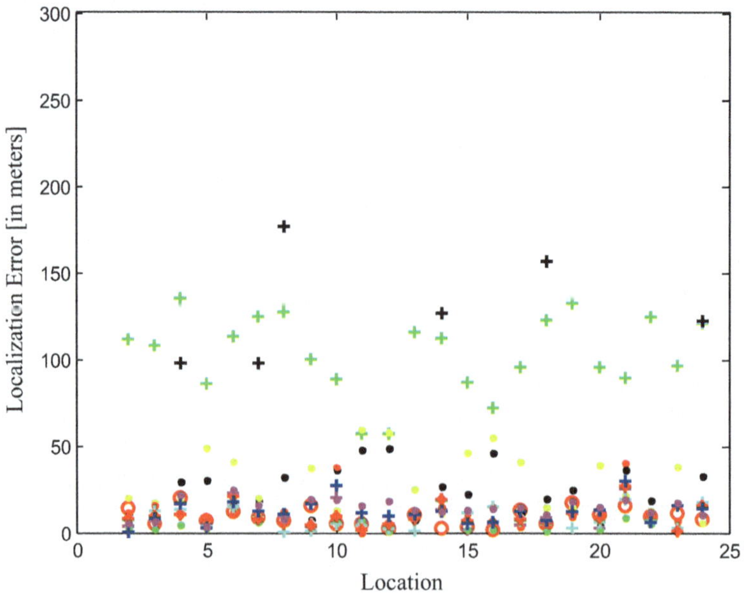

Fig. 7.10 Average localization error with trilateration and FNNN with various training functions with variance of measurement noise = 4 dBm (Case I)

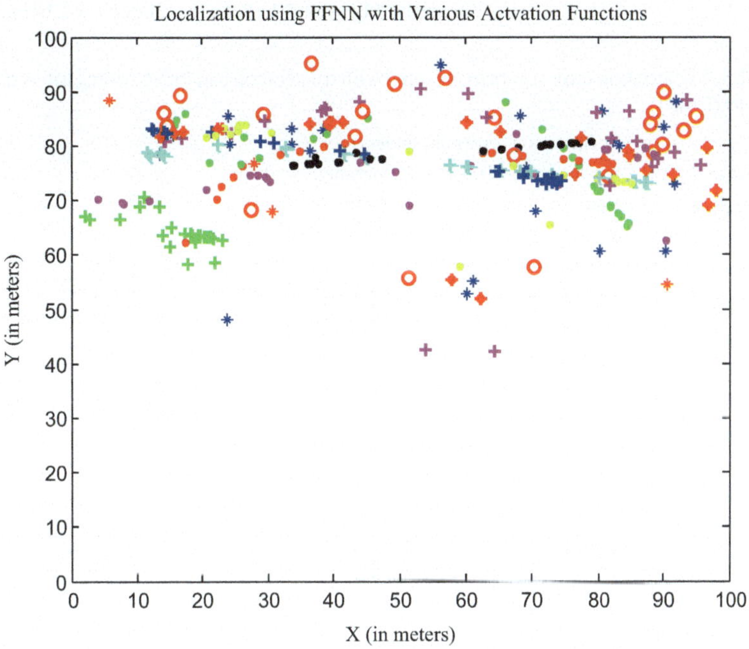

Fig. 7.11 Localization using trilateration and FNNN with various training functions with variance of measurement noise = 5 dBm (Case I)

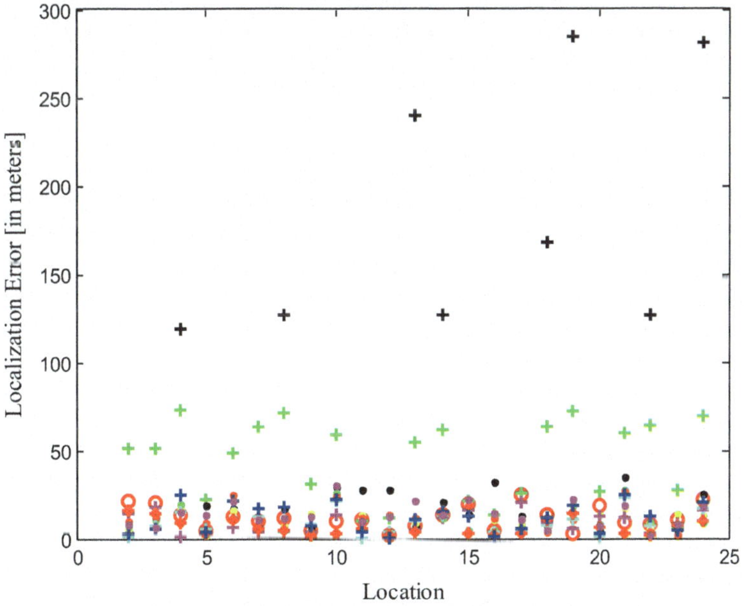

Fig. 7.12 Average localization error with trilateration and FNNN with various training functions with variance of measurement noise = 5 dBm (Case I)

Table 7.5 Localization errors (numeric results) with trilateration and FFNN with various training functions (Case I)

	Localization errors (in meters)		
Algorithm	Variance = 3 dBm	Variance = 4 dBm	Variance = 5 dBm
Trilateration	146	159	167
LM	9.5037	10.3631	12.5560
GDA	9.3820	31.3192	11.6953
GDX	11.2231	24.2291	16.6293
GD	34.3489	101.7567	45.0641
CGB	19.8868	7.8188	11.5441
CGF	11.5619	11.3910	9.8494
CGP	8.9915	12.7732	13.6693
RP	10.9167	9.1506	10.9545
SCG	11.0890	13.4659	13.3057
OSS	10.3161	11.0642	8.2845
BFG	9.1781	15.9339	14.4390

By observing the simulation results (Figs. 7.7, 7.8, 7.9, 7.10, 7.11, and 7.12) and the numeric result (Table 7.6), it can be easily concluded that the localization performance of the FFNT algorithm with few training functions is superior to that with the rest of the other functions. The major research findings of Case I are given below:

1. As the noise variance increased (from 0 dBm to 5 dBm), the average localization error increases drastically with trilateration as against FFNT trained with any other function (see Table 7.6). Thus, trilateration-based L&T approach cannot be recommended for applications, wherein the noise uncertainty is unpredictable.
2. The L&T accuracy for training function GD is worst. The L&T accuracy for GDA, CGF, GDX, CGP, and CGB are inconsistent.
3. The training function LM yields fairly better localization performance for almost all the variance cases. Thus, it may be recommended for the RSSI-based L&T problems.

- ***Case II:** Comparison of trilateration, GRNN, RBFNN, FFNT, and MLP architectures in the context of target L&T.*

The performance comparison of the trilateration, RBFNN, GRNN, FFNT, and MLP architectures (network trained) is depicted in Figs. 7.13, 7.14, 7.15, 7.16, 7.17, and 7.18 for all the RSSI noise variance cases. The numeric comparison results of the average localization errors are given in Table 7.7. In all of these results, the following color combinations are used:

Red marker "∗"—location of anchor nodes.
Blue marker "*"—actual location of unknown nodes.
Black marker "+"—estimated locations with trilateration.
Red marker "o"—estimated locations with GRNN.
Green marker "+"—estimated locations with RBFN.

Table 7.6 Localization errors with trilateration, GRNN, RBFN, MLP, and FNNT for variance = 3 dBm (Case II)

Target location to be estimated	Estimation with trilateration	Estimation with GRNN	Estimation with RBFN	Estimation with MLP	Estimation with FFNN
(23.86, 48.17)	(123.87, 148.12)	(24.86, 49.18)	(33.86, 48.17)	(33.86, 48.17)	(43.66, 58.37)
(69.16, 75.90)	(169.17, 175.93)	(68.16, 75.90)	(79.16, 75.90)	(69.16, 75.90)	(68.33, 65.90)
(70.52, 68.05)	(170.52, 168.00)	(72.52, 68.05)	(90.54, 68.05)	(40.52, 68.05)	(80.52, 78.05)
(91.95, 88.10)	(191.97, 188.10)	(93.95, 90.10)	(91.95, 48.10)	(41.95, 88.10)	(81.23, 98.30)
(24.09, 85.62)	(124.09, 185.64)	(26.09, 85.62)	(44.09, 85.62)	(54.09, 85.62)	(54.11, 55.62)
(56.26, 94.75)	(156.26, 194.75)	(58.26, 94.75)	(76.55, 94.75)	(56.26, 94.75)	(76.26, 74.35)
(80.72, 86.36)	(180.72, 186.37)	(84.72, 86.36)	(90.72, 46.36)	(80.72, 86.36)	(50.72, 76.36)
(91.83, 72.96)	(191.88, 172.96)	(94.83, 74.96)	(91.83, 72.96)	(81.83, 72.96)	(91.81, 62.96)
(38.53, 82.95)	(138.53, 182.95)	(38.53, 82.95)	(58.53, 82.95)	(38.53, 82.95)	(38.53, 82.35)
(61.09, 55.00)	(161.08, 155.00)	(62.09, 55.00)	(81.55, 45.00)	(71.09, 55.00)	(81.09, 55.00)
(13.48, 78.54)	(113.48, 178.54)	(15.48, 78.54)	(33.48, 78.54)	(23.48, 78.54)	(23.49, 68.54)
(13.53, 82.42)	(113.53, 182.42)	(16.53, 82.42)	(43.66, 82.42)	(13.53, 82.42)	(13.53, 82.32)
(68.27, 85.39)	(168.27, 185.39)	(64.27, 85.39)	(88.27, 85.39)	(68.27, 85.39)	(68.27, 55.39)
(80.21, 60.70)	(180.24, 160.70)	(84.21, 60.70)	(90.21, 40.70)	(90.21, 60.70)	(20.20, 70.70)
(24.24, 80.14)	(124.24, 180.14)	(26.24, 82.14)	(64.24, 80.14)	(34.24, 80.14)	(21.21, 80.31)
(15.57, 81.90)	(115.54, 181.92)	(17.57, 81.90)	(55.57, 81.90)	(15.57, 81.90)	(15.57, 85.90)
(33.81, 83.24)	(133.81, 183.24)	(35.81, 83.24)	(33.81, 43.24)	(33.81, 83.24)	(33.81, 89.34)
(82.40, 80.80)	(182.44, 180.83)	(86.40, 80.80)	(42.40, 80.80)	(82.40, 80.80)	(62.40, 80.80)
(90.22, 83.51)	(190.22, 183.53)	(96.22, 83.51)	(40.33, 83.51)	(80.22, 83.51)	(80.11, 83.31)
(33.79, 80.30)	(133.76, 180.30)	(37.79, 80.30)	(43.79, 80.30)	(33.79, 80.30)	(33.79, 87.30)
(60.28, 52.66)	(160.28, 152.66)	(67.28, 52.66)	(60.28, 42.66)	(40.28, 52.66)	(40.67, 52.66)
(83.18, 80.03)	(183.18, 180.33)	(84.18, 80.03)	(63.18, 80.03)	(84.18, 80.03)	(83.18, 89.03)
(36.41, 79.04)	(136.41, 179.04)	(38.41, 79.04)	(36.47, 79.04)	(36.41, 79.04)	(56.41, 79.34)
(90.45, 60.71)	(190.47, 160.71)	(93.45, 60.31)	(60.45, 30.21)	(77.35, 60.71)	(80.24, 69.31)

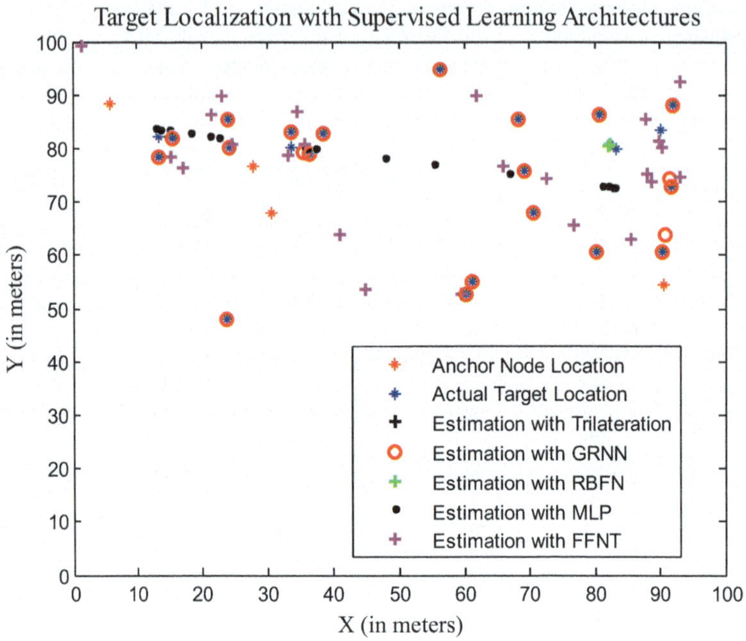

Fig. 7.13 Localization using trilateration, GRNN, RBFN, MLP, and FNNT with variance of measurement noise = 3 dBm (Case II)

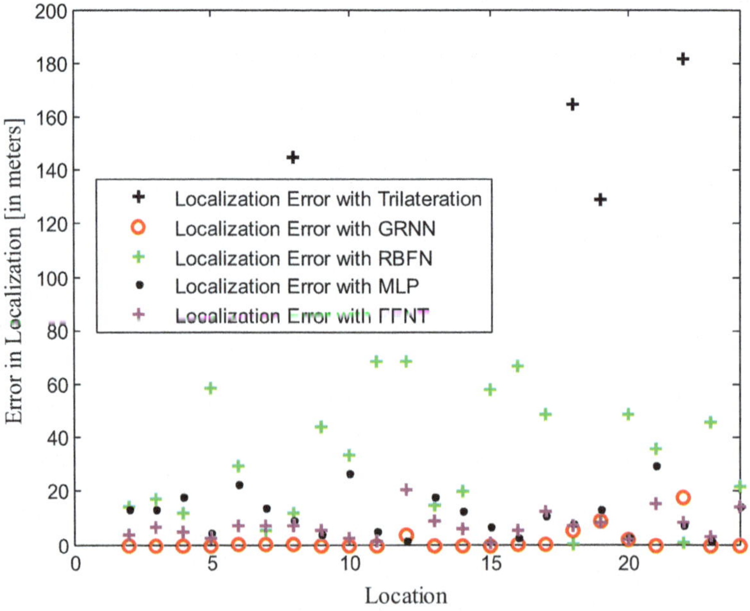

Fig. 7.14 Average localization error with trilateration, GRNN, RBFN, MLP, and FNNT with variance of measurement noise = 3 dBm (Case II)

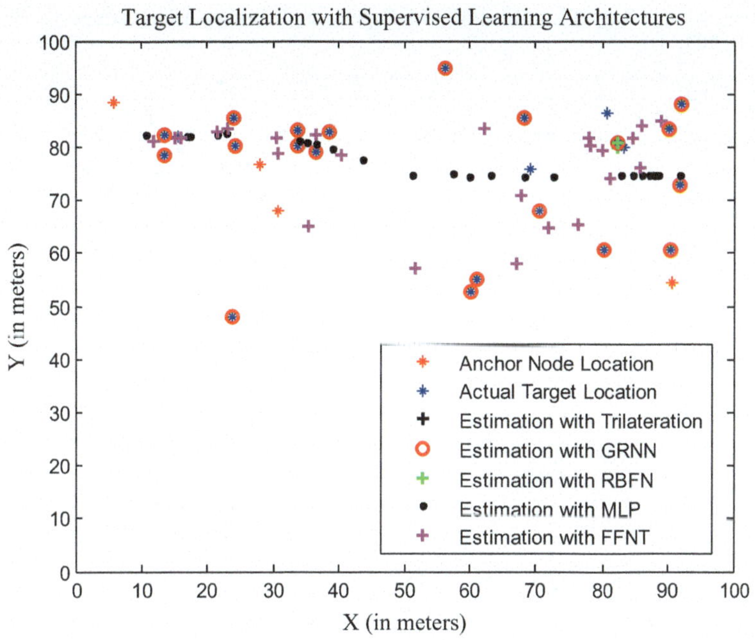

Fig. 7.15 Localization using trilateration, GRNN, RBFN, MLP, and FNNT with variance of measurement noise = 4 dBm (Case II)

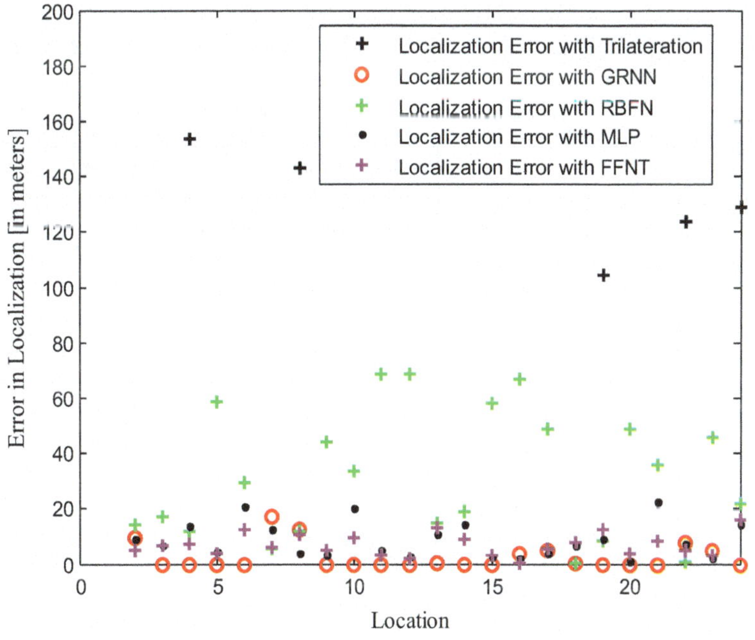

Fig. 7.16 Average localization error with trilateration, GRNN, RBFN, MLP, and FNNT with variance of measurement noise = 4 dBm (Case II)

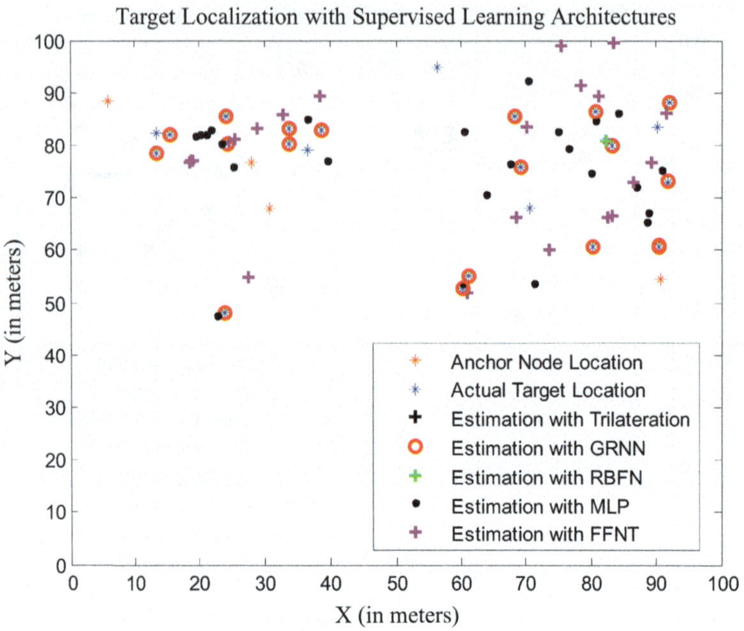

Fig. 7.17 Localization using trilateration, GRNN, RBFN, MLP, and FNNT with variance of measurement noise = 5 dBm (Case II)

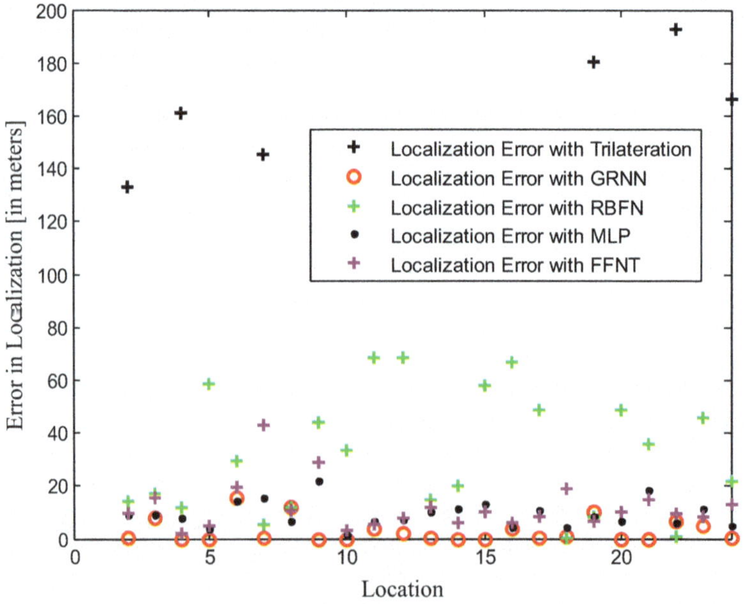

Fig. 7.18 Average localization error with trilateration, GRNN, RBFN, MLP, and FNNT with variance of measurement noise = 5 dBm (Case II)

Table 7.7 Localization errors (numeric comparison) with trilateration, GRNN, RBFN, MLP, and FNNT (Case II)

Average localization error (in meters)			
Algorithm	Variance =3	Variance =4	Variance =5
Trilateration	160	162	165
FNNN	7.7987	7.5275	11.8115
RBFN	33.4080	33.3855	33.4245
GRNN	1.6129	2.5414	2.8646
MLP	12.4832	9.7424	8.9789

Black marker "."—estimated locations with MLP.

Orange marker "+"—estimated locations with FFNT.

From Table 7.6 one can very easily claim that there is a very slight deviation in the estimated location from the actual target location with the GRNN as compared to the trilateration and the rest of other architectures. Table 7.6 also reveals that the RBFN shows the worst estimation performance as compared to all the rest of the architectures for the considered environment. In some instances the localization performance with MLP is superior to others, while at the remaining instances the FFNT is superior to other implementations. It is believed that with changes in the dimension of the RSSI vector, size of the training database, and number of neurons in the hidden layer, one may achieve different localization results.

7.4 Conclusion

This chapter discusses the applications of supervised learning architectures such as FFNT, RBFNN, GRNN, and MLP to the problem of target L&T. All of these architectures are trained with RSSI measurements and the corresponding 2-D locations of the mobile target in off-line phase. Two cases are considered during simulation experiments. In Case I, the impact of various training functions on the FFNT-based target L&T system is analyzed in the context of localization accuracy, whereas in Case II, the target localization performance comparison is made between the traditional trilateration-, FFNT-, MLP-, GRNN-, and RBFN-based target L&T system. In order to accomplish fair comparison in all of the supervised learning architectures in both the cases, all of them are fed with four RSSI measurements, and they are supposed to estimate 2-D target location, corresponding to this input vector of four RSSI measurements. In both of the cases locations of anchor nodes as well as target locations to be estimated are kept fixed. The detailed research findings at the end of each of the simulation cases are discussed.

MATLAB Code for Cases I and II

MATLAB Code for Case I

```
%%%%%%%% main.m %%%%%%

%% WSN Deployment Setting Parameters
clearall
closeall
clc

prompt = 'Enter Frequency in MHz : ';
   Freq = input(prompt);
   Freq = Freq*1000000;

anchorLoc  = [5.93 88.50; 27.92  76.89; 30.68 68.03; 90.75 54.45];

targetLoc = [ 23.86 48.17; 69.16 75.90; 70.52 68.05; 91.95 88.10; 24.09 85.62; 56.26 94.75;
             80.72 86.36; 91.83 72.96; 38.53 82.95; 61.09 55.00; 13.48 78.54; 13.53 82.42;
             68.27 85.39; 80.21 60.70; 24.24 80.14; 15.57 81.90; 33.81 83.24; 82.40 80.80;
             90.22 83.51; 33.79 80.30; 60.28 52.66; 83.18 80.03; 36.41 79.04; 90.45 60.71];

%%%%%%%%%%%% Training Set for FFNN %%%%%%%%%%%%%%%%

RSS_Input  =  [-18.3288613120487,-4.67428444087499,-0.964045513022431,7.57384665182379,-
49.6179429670664,-11.1608291330407,3.41579702813736,4.02897222371426,-
27.0381904713468,0.653628765631545,-64.3463829643857,-72.3601686545670,-
5.03706842385432,3.24293937753768,-43.9544151850606,-67.7661092893538,-
33.7325788696963,0.765613593731738,0.228152864937741,-36.4488914396046,-
1.94267658857350,-0.0951575163411000,-31.3751011668895,5.68595216292656;-
29.8913164659770,-19.5199778478546,-15.5112990033693,-0.746238720943790,-
69.6978786517952,-24.7749824239484,-8.51565108224177,-5.10052020486471,-
62.1398461436167,-17.8057754009193,-54.0669440657066,-54.4809469444066,-
22.7477918146961,-5.23068524545026,-97.0131882992456,-64.3232183767238,-
78.1121213377313,-9.57281580095875,-6.02888004970743,-86.7354520848503,-
17.9366791314333,-10.5879222328873,-78.9911095429087,-4.23699892275761;-
46.9280995126409,-27.0142861556619,-18.2414258118235,4.83660485995008,-
50.5047822844662,-27.0107302653903,-12.8326108403980,-7.40486648364890,-
55.8828708245475,-32.3124366366187,-43.2237831535616,-49.4804736587852,-
25.8576505384169,-14.9478273779381,-63.1861032502966,-46.6976999807710,-
58.2358646652669,-8.18490742509330,-5.60963014163823,-63.8689648134508,-
27.6186257047612,-10.6563791796436,-63.3311467652521,-7.23059508971731;-
26.6799524140441,-45.8212883548265,-51.6157532570173,-9.39507906412217,-
12.5021083347264,-18.0667197448536,-23.8936722245174,-25.2469317055969,-
23.3543929151763,-172.113685391184,-10.9596045673222,-10.4917285436349,-
26.4376611687186,-42.6338151474468,-16.8873312185192,-14.9098863894250,-
19.0728092264533,-25.7573658859847,-20.8358849454565,-19.1657838399446,-
140.561929589991,-24.7580613967916,-27.5448724632910,-28.6697339921816];

   Target = targetLoc';
   spread=1;
```

```
% set early stopping parameters
net.divideParam.trainRatio = 0.7;% training set [%]
net.divideParam.valRatio = 0.15; % validation set [%]
net.divideParam.testRatio = 0.15; % test set [%]
net.trainParam.epochs = 100;

%%*********BPNN-LM*************************
net_LM = feedforwardnet(3,'trainlm');
net_LM = train(net_LM,RSS_Input,Target);

net_GD = feedforwardnet(3,'traingd');
net_GD = train(net_GD,RSS_Input,Target);

net_GDX = feedforwardnet(3,'traingdx');
net_GDX = train(net_GDX,RSS_Input,Target);

net_GDA = feedforwardnet(3,'traingda');
net_GDA = train(net_GDA,RSS_Input,Target);

net_CGF = feedforwardnet(3,'traincgf');
net_CGF = train(net_CGF,RSS_Input,Target);

net_CGB = feedforwardnet(3,'traincgb');
net_CGB = train(net_CGB,RSS_Input,Target);

net_CGP = feedforwardnet(3,'traincgp');
net_CGP = train(net_CGP,RSS_Input,Target);

net_RP = feedforwardnet(3,'trainrp');
net_RP = train(net_RP,RSS_Input,Target);

net_OSS = feedforwardnet(3,'trainoss');
net_OSS = train(net_OSS,RSS_Input,Target);

net_SCG = feedforwardnet(3,'trainscg')
net_SCG = train(net_SCG,RSS_Input,Target);

net_BFG = feedforwardnet(3,'trainbfg');
net_BFG = train(net_BFG,RSS_Input,Target);


%show anchor Locations
  f1 = figure(1);
plot(anchorLoc(:,1),anchorLoc(:,2),'r*');
holdon
plot(targetLoc(:,1),targetLoc(:,2),'b*');
ylabel('Y (in meters)','FontName','Times','Fontsize',14,'LineWidth',2);
xlabel('X (in meters)','FontName','Times','Fontsize',14,'LineWidth',2);
title('Localization using FFNN with Various Actvation Functions','FontName','Times',
'Fontsize',12,'LineWidth',2);
```

```
  axis([0 100 0 100])
  holdon


no_of_positions =24;
Avg_LocError_Trilateration =0; Avg_LocError_LM = 0; Avg_LocError_GD = 0;
Avg_LocError_GDX = 0; Avg_LocError_GDA = 0; Avg_LocError_CGF = 0;
Avg_LocError_CGB = 0; Avg_LocError_CGP = 0; Avg_LocError_RP = 0;
Avg_LocError_OSS = 0;Avg_LocError_SCG = 0; Avg_LocError_BFG = 0;

  % Calculate reference RSSI at d0 = 1 meter using Free Space Path Loss Model
    d0=1;
    Pr0 = RSSI_friss(d0,Freq);

    d_test = 60;

    Pr = RSSI_friss(d_test,Freq);

  %Calculation of Path Loss Exponent :

  n = -(Pr + Pr0)/(10*log(d_test));

  % Generating trajectory for the mobile node

  for t = 1:no_of_positions
  P(t)=targetLoc(t,1);
  Q(t)=targetLoc(t,2);
  end

  for t = 1:no_of_positions

      Actual_Target_Location = targetLoc(t,:)

  % Actual Distances from Anchors required to generate RSSI Values
      d1 = sqrt((5.93-P(t))^2+ (88.50-Q(t))^2);
      d2 = sqrt((27.92-P(t))^2+ (76.89-Q(t))^2);
      d3 = sqrt((30.68-P(t))^2+ (68.03-Q(t))^2);
      d4 = sqrt((60.75-P(t))^2+ (54.45-Q(t))^2);


  % Generate RSSI Values according to OFPEDM Path Loss Model

      RSS = lognormalshadowing_4(n,d1,d2,d3,d4,Pr0);
      RSS_s = sort(RSS);


  % Application of Trilateration based L&T Algorithm:

      Trilateration_Estimated_Loc = trilateration_4(RSS,RSS_s,Pr0,n )
```

```
        X_T = Trilateration_Estimated_Loc(1);
        Trilateration_X(t)=Trilateration_Estimated_Loc(1);

    Y_T = Trilateration_Estimated_Loc(2);
        Trilateration_Y(t)=Trilateration_Estimated_Loc(2);

    plot(X_T,Y_T,'k+','LineWidth',2)
    holdon

        RSS_1(t)= RSS(1);
        RSS_2(t)= RSS(2);
        RSS_3(t)= RSS(3);
        RSS_4(t)= RSS(4);


% Online Estimation Phase: Application Set of four random RSSI values to Considered
   Supervised Learning Architectures:

        RSS = [RSS_1(t), RSS_2(t), RSS_3(t), RSS_4(t)];
        RSS_new_vector = RSS.';

        LM_Estimated_Loc= net_LM(RSS_new_vector)
        LM_X(t)= LM_Estimated_Loc(1);
        LM_Y(t)= LM_Estimated_Loc(2);
    plot(LM_Estimated_Loc(1),LM_Estimated_Loc(2),'ro','LineWidth',2)

        GD_Estimated_Loc= net_GD(RSS_new_vector)
        GD_X(t)= GD_Estimated_Loc(1);
        GD_Y(t)= GD_Estimated_Loc(2);
    plot(GD_Estimated_Loc(1), GD_Estimated_Loc(2),'g+','LineWidth',2)


        GDX_Estimated_Loc= net_GDX(RSS_new_vector)
        GDX_X(t)= GDX_Estimated_Loc(1);
        GDX_Y(t)= GDX_Estimated_Loc(2);
    plot(GDX_Estimated_Loc(1), GDX_Estimated_Loc(2),'k.','LineWidth',2)

        GDA_Estimated_Loc= net_GDA(RSS_new_vector)
        GDA_X(t)= GDA_Estimated_Loc(1);
        GDA_Y(t)= GDA_Estimated_Loc(2);
    plot(GDA_Estimated_Loc(1), GDA_Estimated_Loc(2),'y.','LineWidth',2)

        CGF_Estimated_Loc= net_CGF(RSS_new_vector)
        CGF_X(t)= CGF_Estimated_Loc(1);
        CGF_Y(t)= CGF_Estimated_Loc(2);
    plot(CGF_Estimated_Loc(1), CGF_Estimated_Loc(2),'m+','LineWidth',2)

        CGB_Estimated_Loc= net_CGB(RSS_new_vector)
        CGB_X(t)= CGB_Estimated_Loc(1);
```

```
    CGB_Y(t)= CGB_Estimated_Loc(2);
plot(CGB_Estimated_Loc(1), CGB_Estimated_Loc(2),'g.','LineWidth',2)

    CGP_Estimated_Loc= net_CGP(RSS_new_vector)
    CGP_X(t)= CGP_Estimated_Loc(1);
    CGP_Y(t)= CGP_Estimated_Loc(2);
plot(CGP_Estimated_Loc(1), CGP_Estimated_Loc(2),'r.','LineWidth',2)

    RP_Estimated_Loc= net_RP(RSS_new_vector)
    RP_X(t)= RP_Estimated_Loc(1);
    RP_Y(t)= RP_Estimated_Loc(2);
plot(RP_Estimated_Loc(1), RP_Estimated_Loc(2),'c+','LineWidth',2)

    OSS_Estimated_Loc= net_OSS(RSS_new_vector)
    OSS_X(t)= OSS_Estimated_Loc(1);
    OSS_Y(t)= OSS_Estimated_Loc(2);
plot(OSS_Estimated_Loc(1), OSS_Estimated_Loc(2),'r*','LineWidth',2)

    SCG_Estimated_Loc= net_SCG(RSS_new_vector)
    SCG_X(t)= SCG_Estimated_Loc(1);
    SCG_Y(t)= SCG_Estimated_Loc(2);
plot(SCG_Estimated_Loc(1), SCG_Estimated_Loc(2),'b+','LineWidth',2)


    BFG_Estimated_Loc= net_BFG(RSS_new_vector)
    BFG_X(t)= BFG_Estimated_Loc(1);
    BFG_Y(t)= BFG_Estimated_Loc(2);
plot(BFG_Estimated_Loc(1), BFG_Estimated_Loc(2),'m.','LineWidth',2)

%%%%% Error Analysis of algorithm

LocError_Trilateration = sqrt((P(t) - Trilateration_Estimated_Loc(1))^2 + (Q(t) -
Trilateration_Estimated_Loc(2))^2)

Trilateration_Error(t) = LocError_Trilateration;
Avg_LocError_Trilateration = Avg_LocError_Trilateration + LocError_Trilateration;

LocError_LM = sqrt((P(t) - LM_Estimated_Loc(1))^2 + (Q(t) - LM_Estimated_Loc(2))^2)
LM_Error(t) = LocError_LM;
Avg_LocError_LM = Avg_LocError_LM + LocError_LM;

LocError_GD = sqrt((P(t) - GD_Estimated_Loc(1))^2 + (Q(t) - GD_Estimated_Loc(2))^2)
GD_Error(t) = LocError_GD;
Avg_LocError_GD = Avg_LocError_GD + LocError_GD;

LocError_GDX = sqrt((P(t) - GDX_Estimated_Loc(1))^2 + (Q(t) -
GDX_Estimated_Loc(2))^2)

GDX_Error(t) = LocError_GDX;
Avg_LocError_GDX = Avg_LocError_GDX + LocError_GDX;
```

```
LocError_GDA = sqrt((P(t) - GDA_Estimated_Loc(1))^2 + (Q(t) -
GDA_Estimated_Loc(2))^2)

GDA_Error(t) = LocError_GDA;
    Avg_LocError_GDA = Avg_LocError_GDA + LocError_GDA;

LocError_CGF = sqrt((P(t) - CGF_Estimated_Loc(1))^2 + (Q(t) -
CGF_Estimated_Loc(2))^2)

    CGF_Error(t) = LocError_CGF;
    Avg_LocError_CGF = Avg_LocError_CGF + LocError_CGF;

LocError_CGB = sqrt((P(t) - CGB_Estimated_Loc(1))^2 + (Q(t) -
CGB_Estimated_Loc(2))^2)
    CGB_Error(t) = LocError_CGB;
    Avg_LocError_CGB = Avg_LocError_CGB + LocError_CGB;

LocError_CGP = sqrt((P(t) - CGP_Estimated_Loc(1))^2 + (Q(t) -
CGP_Estimated_Loc(2))^2)
    CGP_Error(t) = LocError_CGP;
    Avg_LocError_CGP = Avg_LocError_CGP + LocError_CGP;

LocError_RP = sqrt((P(t) - RP_Estimated_Loc(1))^2 + (Q(t) - RP_Estimated_Loc(2))^2)
    RP_Error(t) = LocError_RP;
    Avg_LocError_RP = Avg_LocError_RP + LocError_RP;

LocError_OSS = sqrt((P(t) - OSS_Estimated_Loc(1))^2 + (Q(t) - OSS_Estimated_Loc(2))^2)
    OSS_Error(t) = LocError_OSS;
    Avg_LocError_OSS = Avg_LocError_OSS + LocError_OSS;

LocError_SCG = sqrt((P(t) - SCG_Estimated_Loc(1))^2 + (Q(t) -
SCG_Estimated_Loc(2))^2)
    SCG_Error(t) = LocError_SCG;
    Avg_LocError_SCG = Avg_LocError_SCG + LocError_SCG;

LocError_BFG = sqrt((P(t) - BFG_Estimated_Loc(1))^2 + (Q(t) -
BFG_Estimated_Loc(2))^2)
    BFG_Error(t) = LocError_BFG;
    Avg_LocError_BFG = Avg_LocError_BFG + LocError_BFG;

  end

legend('Anchor Node Location','Actual Target Track','Estimation with
Trilateration','Estimation with trainlm','Estimation with traingd','Estimation with
traingdx','Estimation with traingda','Estimation with traincgf','Estimation with
traincgb','Estimation with traincgp','Estimation with trainrp','Estimation with
trainoss','Estimation with trainscg','Estimation with trainbfg','Location','BestOutside')
```

```
Avg_LocError_Trilateration = Avg_LocError_Trilateration/no_of_positions
Avg_LocError_LM = Avg_LocError_LM/no_of_positions
Avg_LocError_GD = Avg_LocError_GD/no_of_positions
Avg_LocError_GDX = Avg_LocError_GDX/no_of_positions
Avg_LocError_GDA = Avg_LocError_GDA/no_of_positions
Avg_LocError_CGF = Avg_LocError_CGF/no_of_positions
Avg_LocError_CGB = Avg_LocError_CGB/no_of_positions
Avg_LocError_CGP = Avg_LocError_CGP/no_of_positions
Avg_LocError_RP = Avg_LocError_RP/no_of_positions
Avg_LocError_OSS = Avg_LocError_OSS/no_of_positions
Avg_LocError_SCG = Avg_LocError_SCG/no_of_positions
Avg_LocError_BFG = Avg_LocError_BFG/no_of_positions

 f2 = figure(2);
for t =1:no_of_positions

plot(t,LM_Error(t),'ro','LineWidth',2)
plot(t,GD_Error(t),'g+','LineWidth',2)
plot(t,GDX_Error(t),'k.','LineWidth',2)
plot(t,GDA_Error(t),'y.','LineWidth',2)
plot(t,CGF_Error(t),'m+','LineWidth',2)
plot(t,CGB_Error(t),'g.','LineWidth',2)
plot(t,CGP_Error(t),'r.','LineWidth',2)
plot(t,RP_Error(t),'c+','LineWidth',2)
plot(t,OSS_Error(t),'r*','LineWidth',2)
plot(t,SCG_Error(t),'b+','LineWidth',2)
plot(t,BFG_Error(t),'m.','LineWidth',2)
plot(t,Trilateration_Error(t),'k+','LineWidth',2)
axis([0 25 0 300])
xlabel('Location','FontName','Times','Fontsize',14,'LineWidth',2)
ylabel('Localization Error [in meters]','FontName','Times','Fontsize',14,'LineWidth',2)
holdon

end

legend('Localization Error with Trilateration','Localization Error with trainlm','Localization
Error with traingd','Localization Error with traingdx','Localization Error with
traingda','Localization Error with traincgf', 'Localization Error with traincgb','Localization
Error with traincgp','Localization Error with trainrp','Localization Error with
trainoss','Localization Error with trainscg','Localization Error with
trainbfg','Location','BestOutside')
```

%%%%%%%% lognormalshadowing_4.m %%%%%%%%%%%%%%%

```
function [ RSS ] = lognormalshadowing_4(n,d1,d2,d3,d4,Pr0 )
%UNTITLED5 Summary of this function goes here
%  Example Generate values from a normal distribution with mean 1 and standard
deviation 2:
```

```
mean = 0;  % standard deviation 0f measurement noise in dbm
variance =3;

        RSS_d1= (10*n* log10(d1) + Pr0) + (mean + variance.*randn);
        RSS_d2= (10*n* log10(d2) + Pr0) + (mean + variance.*randn);
        RSS_d3= (10*n* log10(d3) + Pr0) + (mean + variance.*randn);
        RSS_d4= (10*n* log10(d4) + Pr0) + (mean + variance.*randn);

        RSS = [ RSS_d1, RSS_d2, RSS_d3, RSS_d4];
end


%%%%%%%%%% Trilateration.m %%%%%%%%%%%%%%%%%%

function [ mobileLoc_est ] = trilateration_4( RSS,RSS_s,Pr0,n )
%UNTITLED6 Summary of this function goes here
%   Detailed explanation goes here

% Select  highest three RSSI Values  & Calculate distances using d = antilog(Pr0-
RSSI)/(10*n)

          d1_est = 10^(-(Pr0+RSS_s(4))/(10*n));
          d2_est = 10^(-(Pr0+RSS_s(3))/(10*n));
          d3_est = 10^(-(Pr0+RSS_s(2))/(10*n));


if (RSS_s(4) == RSS(1))
            X1 = 5.93; Y1 = 88.50;
elseif (RSS_s(4) == RSS(2))
            X1 = 27.92; Y1 = 76.89;
elseif (RSS_s(4) == RSS(3))
            X1 = 30.68; Y1 = 68.03;
elseif (RSS_s(4) == RSS(4))
            X1 = 90.75; Y1 = 54.45;
end


if (RSS_s(3) == RSS(1))
            X2 = 5.93; Y2 = 88.50;
elseif (RSS_s(3) == RSS(2))
            X2 = 27.92; Y2 = 76.89;
elseif (RSS_s(3) == RSS(3))
            X2 = 30.68; Y2 = 68.03;
elseif (RSS_s(3) == RSS(4))
            X2 = 90.75; Y2 = 54.45;
end


if (RSS_s(2) == RSS(1))
            X3 = 5.93; Y3 = 88.50;
```

```
elseif (RSS_s(2) == RSS(2))
        X3 = 27.92; Y3 = 76.89;
elseif (RSS_s(2) == RSS(3))
        X3 = 30.68; Y3 = 68.03;
elseif (RSS_s(2) == RSS(4))
        X3 = 90.75; Y3 = 54.45;
end


    A = X1^2 + Y1^2 - d1_est^2;
    B = X2^2 + Y2^2 - d2_est^2;
    C = X3^2 + Y3^2 - d3_est^2;

    X32 = (X3 - X2);
    Y32 = (Y3 - Y2);
    X21 = (X2 - X1);
    Y21 = (Y2 - Y1);
    X13 = (X1 - X3);
    Y13 = (Y1 - Y3);

X_T = (A*Y32 + B*Y13 + C*Y21) / (2*( X1*Y32 + X2*Y13 + X3*Y21));
Y_T = (A*X32 + B*X13 + C*X21) / (2*( Y1*X32 + Y2*X13 + Y3*X21));

    mobileLoc_est = [ X_T, Y_T];

end
```

%%%%%%%%%% RSSI_friss.m %%%%%%%%%%%%%%%%%%%%

```
function [ Pr ] = RSSI_friss( d,Freq )

C=3e8;              %LightSpeed

  Freq = Freq*1000000;
  Zigbee=915.0e6;%hz
  TXAntennaGain=1;%db
  RXAntennaGain=1;%db

  PTx=0.001;%watt

    Wavelength=C/Freq;
    PTxdBm=10*log10(PTx*1000);
    M = Wavelength / (4 * pi * d);

Pr=PTxdBm + TXAntennaGain + RXAntennaGain- (20*log10(1/M));

end
```

MATLAB Code for Case II

```
%%%%%%%%%%%% Main.m %%%%%%%%%%%%%%%%%%%

%% WSN Deployment Setting Parameters
clearall
closeall
clc

prompt = 'Enter Frequency in MHz : ';
  Freq = input(prompt);
  Freq = Freq*1000000;

anchorLoc  = [5.93 88.50; 27.92  76.89; 30.68 68.03; 90.75 54.45];

targetLoc= [ 23.86 48.17; 69.16 75.90; 70.52 68.05; 91.95 88.10; 24.09 85.62; 56.26 94.75;
             80.72 86.36; 91.83 72.96; 38.53 82.95; 61.09 55.00; 13.48 78.54; 13.53 82.42;
             68.27 85.39; 80.21 60.70; 24.24 80.14; 15.57 81.90; 33.81 83.24; 82.40 80.80;
             90.22 83.51; 33.79 80.30; 60.28 52.66; 83.18 80.03; 36.41 79.04; 90.45 60.71];


%%%%%%%%%%%% Training of GRNN %%%%%%%%%%%%

RSS_Input  =  [-18.3288613120487,-4.67428444087499,-0.964045513022431,7.57384665182379,-
49.6179429670664,-11.1608291330407,3.41579702813736,4.02897222371426,-
27.0381904713468,0.653628765631545,-64.3463829643857,-72.3601686545670,-
5.03706842385432,3.24293937753768,-43.9544151850606,-67.7661092893538,-
33.7325788696963,0.765613593731738,0.228152864937741,-36.4488914396046,-
1.94267658857350,-0.0951575163411000,-31.3751011668895,5.68595216292656;-
29.8913164659770,-19.5199778478546,-15.5112990033693,-0.746238720943790,-
69.6978786517952,-24.7749824239484,-8.51565108224177,-5.10052020486471,-
62.1398461436167,-17.8057754009193,-54.0669440657066,-54.4809469444066,-
22.7477918146961,-5.23068524545026,-97.0131882992456,-64.3232183767238,-
78.1121213377313,-9.57281580095875,-6.02888004970743,-86.7354520848503,-
17.9366791314333,-10.5879222328873,-78.9911095129087, 4.23699892275761;-
46.9280995126409,-27.0142861556619,-18.2414258118235,4.83660485995008,-
50.5047822844662,-27.0107302653903,-12.8326108403980,-7.40486648364890,-
55.8828708245475,-32.3124366366187,-43.2237831535616,-49.4804736587852,-
25.8576505384169,-14.9478273779381,-63.4861032502966,-46.6976999807710,-
58.2358646652669,-8.18490742509330, 5.60963014163823,-63.8689648134508,-
27.6186257047612,-10.6563791796436,-63.3311467652521,-7.23059508971731;-
26.6799524140441,-45.8212883548265,-51.6157532570173,-9.39507906412217,-
12.5021083347264,-18.0667197448536,-23.8936722245174,-25.2469317055969,-
23.3543929151763,-172.113685391184,-10.9596045673222,-10.4917285436349,-
26.4376611687186,-42.6338151474468,-16.8873312185192,-14.9098863894250,-
19.0728092264533,-25.7573658859847,-20.8358849454565,-19.1657838399446,-
140.561929589991,-24.7580613967916,-27.5448724632910,-28.6697339921816];

  Target = targetLoc';
```

```
spread=1;

net_Loc_est = newgrnn(RSS_Input,Target,spread);
view(net_Loc_est)

%%%%%%%%%%%%%%%%% Training of Exact RBFN %%%%%%%%%%
K = 80;
goal = 0;     % performance goal (SSE)
Ki = 25;      % number of neurons to add between displays
net_RBFN = newrb(RSS_Input,Target,goal,spread,K, Ki)

%%%%%%%%%%%%%%%%%% Training of MLP %%%%%%%%%%%%%%%%%
net_MLP = feedforwardnet([5 3]);
net_MLP1 = train(net_MLP,RSS_Input,Target);
% set early stopping parameters
net.divideParam.trainRatio = 0.7; % training set [%]
net.divideParam.valRatio = 0.15; % validation set [%]
net.divideParam.testRatio = 0.15; % test set [%]
net.trainParam.epochs = 100;

%%%%%%%%%%%%%%%% Training of FFNN %%%%%%%%%%%%%
net_LM = feedforwardnet(3,'trainlm');
net_LM1 = train(net_LM,RSS_Input,Target);


%show anchor Locations
   f1 = figure(1);
plot(anchorLoc(:,1),anchorLoc(:,2),'r*');
holdon
plot(targetLoc(:,1),targetLoc(:,2),'b*');
ylabel('Y (in meters)','FontName','Times','Fontsize',14,'LineWidth',2);
xlabel('X (in meters)','FontName','Times','Fontsize',14,'LineWidth',2);

title('Target Localization with Supervised Learning Architectures','FontName',
'Times','Fontsize',12,'LineWidth',2);
axis([0 100 0 100])
holdon


no_of_positions =24;
Avg_LocError_Trilateration =0; Avg_LocError_GRNN = 0; Avg_LocError RBFN = 0;
Avg_LocError_MLP = 0, Avg_LocError_FFNN = 0;

% Calculate reference RSSI at d0 = 1 meter using Free Space Path Loss Model

d0=1;                          % in Meters
   Pr0 = RSSI_friss(d0,Freq);

   d_test = 60;
   Pr = RSSI_friss(d_test,Freq);
```

```
%Calculation of Path Loss Exponent:
n = -(Pr + Pr0)/(10*log(d_test));

% Generating trajectory for the mobile node
for t = 1:no_of_positions
P(t)=targetLoc(t,1);
Q(t)=targetLoc(t,2);
end

for t = 1:no_of_positions

    Actual_Target_Location = targetLoc(t,:)
% Actual Distances from Anchors required to generate RSSI Values
    d1 = sqrt((5.93-P(t))^2+ (88.50-Q(t))^2);
    d2 = sqrt((27.92-P(t))^2+ (76.89-Q(t))^2);
    d3 = sqrt((30.68-P(t))^2+ (68.03-Q(t))^2);
    d4 = sqrt((60.75-P(t))^2+ (54.45-Q(t))^2);

% Generate RSSI Values according to OFPEDM Path Loss Model

    RSS = lognormalshadowing_4(n,d1,d2,d3,d4,Pr0);
    RSS_s = sort(RSS);

% Application of Trilateration Algorithm:

    Trilateration_Estimated_Loc = trilateration_4(RSS,RSS_s,Pr0,n )

    X_T = Trilateration_Estimated_Loc(1);
    Trilateration_X(t)=Trilateration_Estimated_Loc(1);
    Y_T = Trilateration_Estimated_Loc(2);
    Trilateration_Y(t)=Trilateration_Estimated_Loc(2);
plot(X_T,Y_T,'k+','LineWidth',2)
holdon

    RSS_1(t)= RSS(1);
    RSS_2(t)= RSS(2);
    RSS_3(t)= RSS(3);
    RSS_4(t)= RSS(4);

% Online Estimation Phase: Application Considered Supervised Learning Architectures:

    RSS = [RSS_1(t), RSS_2(t), RSS_3(t), RSS_4(t)];
    RSS_new_vector = RSS.';
    GRNN_Estimated_Loc = sim(net_Loc_est,RSS_new_vector)
    GRNN_X(t)= GRNN_Estimated_Loc(1);
    GRNN_Y(t)= GRNN_Estimated_Loc(2);
plot(GRNN_Estimated_Loc(1),GRNN_Estimated_Loc(2),'ro','LineWidth',2)

    RBFN_Estimated_Loc= net_RBFN(RSS_new_vector)
```

```
        RBFN_X(t)= RBFN_Estimated_Loc(1);
        RBFN_Y(t)= RBFN_Estimated_Loc(2);
    plot(RBFN_Estimated_Loc(1), RBFN_Estimated_Loc(2),'g+','LineWidth',2)

        MLP_Estimated_Loc= net_MLP1(RSS_new_vector)
        MLP_X(t)= MLP_Estimated_Loc(1);
        MLP_Y(t)= MLP_Estimated_Loc(2);
    plot(MLP_Estimated_Loc(1), MLP_Estimated_Loc(2),'k.','LineWidth',2)

        FFNN_Estimated_Loc= net_LM1(RSS_new_vector)
        FFNN_X(t)= FFNN_Estimated_Loc(1);
        FFNN_Y(t)= FFNN_Estimated_Loc(2);
    plot(FFNN_Estimated_Loc(1), FFNN_Estimated_Loc(2),'m+','LineWidth',2)

    %%%%% Error Analysis of considered algorithms

LocError_Trilateration = sqrt((P(t) - Trilateration_Estimated_Loc(1))^2 + (Q(t) -
Trilateration_Estimated_Loc(2))^2)

        Trilateration_Error(t)= LocError_Trilateration;
        Avg_LocError_Trilateration = Avg_LocError_Trilateration + LocError_Trilateration;

LocError_GRNN = sqrt((P(t) - GRNN_Estimated_Loc(1))^2 + (Q(t) -
GRNN_Estimated_Loc(2))^2)

        GRNN_Error(t) = LocError_GRNN;
        Avg_LocError_GRNN = Avg_LocError_GRNN + LocError_GRNN;

LocError_RBFN = sqrt((P(t) - RBFN_Estimated_Loc(1))^2 + (Q(t) -
RBFN_Estimated_Loc(2))^2)
        RBFN_Error(t) = LocError_RBFN;
        Avg_LocError_RBFN = Avg_LocError_RBFN + LocError_RBFN;

LocError_MLP = sqrt((P(t) - MLP_Estimated_Loc(1))^2 + (Q(t) -
MLP_Estimated_Loc(2))^2)
        MLP_Error(t) = LocError_MLP;
        Avg_LocError_MLP = Avg_LocError_MLP + LocError_MLP;

LocError_FFNN = sqrt((P(t) - FFNN_Estimated_Loc(1))^2 + (Q(t) -
FFNN_Estimated_Loc(2))^2)
        FFNN_Error(t) = LocError_FFNN;
        Avg_LocError_FFNN = Avg_LocError_FFNN + LocError_FFNN;
    end

legend('Anchor Node Location','Actual Target Location','Estimation with
Trilateration','Estimation with GRNN','Estimation with RBFN','Estimation with
MLP','Estimation with FFNT','Location','SouthWest')

      Avg_LocError_Trilateration = Avg_LocError_Trilateration/no_of_positions
      Avg_LocError_GRNN = Avg_LocError_GRNN/no_of_positions
```

```
    Avg_LocError_RBFN = Avg_LocError_RBFN/no_of_positions
    Avg_LocError_MLP = Avg_LocError_MLP/no_of_positions
    Avg_LocError_FFNN = Avg_LocError_FFNN/no_of_positions

    f2 = figure(2);
  for t =1:no_of_positions


  plot(t,GRNN_Error(t),'ro','LineWidth',2)
  plot(t,RBFN_Error(t),'g+','LineWidth',2)
  plot(t,MLP_Error(t),'k.','LineWidth',2)
  plot(t,FFNN_Error(t),'m+','LineWidth',2)
  plot(t,Trilateration_Error(t),'k+','LineWidth',2)
  axis([0 24 0 200])
  xlabel('Location','FontName','Times','Fontsize',14,'LineWidth',2)
  ylabel('Error in Localization [in meters]','FontName','Times','Fontsize',14,'LineWidth',2)
  holdon

  end

legend('Localization Error with Trilateration','Localization Error with GRNN','Localization
Error with RBFN','Localization Error with MLP','Localization Error with
FFNT','Location','NorthEast')
```

References

1. Y. Zhang, S. Wang, G. Ji, P. Phillips, Fruit classification using computer vision and feedforward neural network. J. Food Eng. **143**, 167–177 (2014). https://doi.org/10.1016/j.jfoodeng.2014.07.001
2. T.K. Gupta, K. Raza, Optimizing deep feedforward neural network architecture: a tabu search based approach. Neural Process. Lett. (3), 51, 2855–2870 (2020). https://doi.org/10.1007/s11063-020-10234-7
3. G.S. Shehu, N. Çetinkaya, Flower pollination–feedforward neural network for load flow forecasting in smart distribution grid. Neural Comput. Appl. **31**(10) (2019). https://doi.org/10.1007/s00521-018-3421-5
4. S. K. Gharghan, R. Nordin, M. Ismail, J. A. Ali, Accurate Wireless Sensor Localization Technique Based on Hybrid PSO-ANN Algorithm for Indoor and Outdoor Track Cycling. IEEE Sensors J. (2016). https://doi.org/10.1109/JSEN.2015.2483745
5. H. Huang, L. Chen, E. Hu, A neural network-based multi-zone modelling approach for predictive control system design in commercial buildings. Energy Build. (2015). https://doi.org/10.1016/j.enbuild.2015.03.045
6. P. Ondruska, I. Posner, Deep tracking: seeing beyond seeing using recurrent neural networks, in *AAAI Conference* (2016). https://doi.org/10.1109/CVPRW.2016.160
7. F. Viani, P. Rocca, G. Oliveri, D. Trinchero, A. Massa, Localization, tracking, and imaging of targets in wireless sensor networks: an invited review. Radio Sci. (2011). https://doi.org/10.1029/2010RS004561
8. W. W. Y. Ng, S. Xu, T. Wang, S. Zhang, C. Nugent, Radial basis function neural network with localized stochastic-sensitive autoencoder for home-based activity recognition. Sensors (Switzerland) **20**(5) (2020). https://doi.org/10.3390/s20051479

9. M. Dua, R. Gupta, M. Khari, R. G. Crespo, Biometric iris recognition using radial basis function neural network. Soft Comput. **23**(22) (2019), https://doi.org/10.1007/s00500-018-03731-4
10. M. A. Mansor, S. Z. M. Jamaludin, M. S. M. Kasihmuddin, S. A. Alzaeemi, M. F. M. Basir, S. Sathasivam, Systematic boolean satisfiability programming in radial basis function neural network. Processes **8**(2) (2020). https://doi.org/10.3390/pr8020214
11. S. H. Bak et al., A study on red tide detection technique by using multi-layer perceptron. Int. J. Grid Distrib. Comput. **11**(9) (2018). https://doi.org/10.14257/ijgdc.2018.11.9.08
12. N. Talebi, A. M. Nasrabadi, I. Mohammad-Rezazadeh, Estimation of effective connectivity using multi-layer perceptron artificial neural network. Cogn. Neurodyn. **12**(1) (2018). https://doi.org/10.1007/s11571-017-9453-1
13. Y. Liu, S. Liu, Y. Wang, F. Lombardi, J. Han, A stochastic computational multi-layer perceptron with backward propagation. IEEE Trans. Comput. **67**(9) (2018). https://doi.org/10.1109/TC.2018.2817237
14. I. Lorencin, N. Anđelić, J. Španjol, Z. Car, Using multi-layer perceptron with Laplacian edge detector for bladder cancer diagnosis. Artif. Intell. Med. **102** (2020). https://doi.org/10.1016/j.artmed.2019.101746
15. E. Volná, M. Kotyrba, Z. K. Oplatková, R. Senkerik, Elliott waves classification by means of neural and pseudo neural networks. Soft Comput. **22**(6) (2018). https://doi.org/10.1007/s00500-016-2097-y
16. S. M. Ghoreishi, A. Hedayati, S. O. Mousavi, Quercetin extraction from Rosa damascena Mill via supercritical CO2: Neural network and adaptive neuro fuzzy interface system modeling and response surface optimization. J. Supercrit. Fluids **112** (2016). https://doi.org/10.1016/j.supflu.2016.02.006
17. M. N. Amar, M. A. Ghriga, H. Ouaer, M. El Amine Ben Seghier, B. T. Pham, P. Ø. Andersen, Modeling viscosity of CO2 at high temperature and pressure conditions. J. Nat. Gas Sci. Eng. **77** (2020). https://doi.org/10.1016/j.jngse.2020.103271
18. A. Ghosh, Comparative study of financial time series prediction by artificial neural network with gradient descent learning. Int. J. Sci. Eng. Res. **3** (2011). arXiv:1111.4930
19. J. Wang, Y. Wen, Y. Gou, Z. Ye, H. Chen, Fractional-order gradient descent learning of BP neural networks with Caputo derivative. Neural Netw. **89** (2017). https://doi.org/10.1016/j.neunet.2017.02.007
20. I. Riadi, A. Wirawan, Sunardi, Network packet classification using neural network based on training function and hidden layer neuron number variation. Int. J. Adv. Comput. Sci. Appl. **8**(6) (2017). https://doi.org/10.14569/ijacsa.2017.080631
21. S.R. Jondhale, R.S. Deshpande, Modified Kalman filtering framework based real time target tracking against environmental dynamicity in wireless sensor networks. Ad Hoc Sens. Wirel. Netw. **40**(1–2), 119–143 (2018)
22. S. R. Jondhale, R. S. Deshpande, GRNN and KF framework based real time target tracking using PSOC BLE and smartphone. Ad Hoc Netw. **84** (2019). https://doi.org/10.1016/j.adhoc.2018.09.017
23. S. R. Jondhale, R. S. Deshpande, Kalman filtering framework-based real time target tracking in wireless sensor networks using generalized regression neural networks. IEEE Sensors J. (2019). https://doi.org/10.1109/JSEN.2018.2873357
24. S. R. Jondhale, R. S. Deshpande, S. M. Walke, A. S. Jondhale, Issues and challenges in RSSI based target localization and tracking in wireless sensor networks, in *2016 International Conference on Automatic Control and Dynamic Optimization Techniques (ICACDOT)* (2017). https://doi.org/10.1109/ICACDOT.2016.7877655

Index

S. R. Jondhale et al., *Received Signal Strength Based Target Localization and Tracking Using Wireless Sensor Networks*, EAI/Springer Innovations in Communication and Computing, https://doi.org/10.1007/978-3-030-74061-0

The manufacturer's authorised representative in the EU is Springer Nature Customer Service Centre GmbH, Europaplatz 3, 69115 Heidelberg, Germany. If you have any concerns regarding our products, please contact ProductSafety@springernature.com

Printed and bound by CPI Group (UK) Ltd, Croydon, CR0 4YY
30/09/2026
02225694-0005